Buying Greenhouse Insurance

Buying Greenhouse Insurance:

The Economic Costs of Carbon Dioxide Emission Limits

Alan S. Manne
Richard G. Richels

The MIT Press
Cambridge, Massachusetts
London, England

Second printing, 1993

©1992 Massachusetts Institute of Technology

This book was set in Palatino by Digital Graphics, Inc. using TEX and printed and bound in the United States of America.

Manne, Alan S. (Alan Sussmann)
    Buying greenhouse insurance : the economic costs of carbon dioxide emission limits / Alan S. Manne, Richard G. Richels.
      p. cm.
    Includes bibliographic references and index.
    ISBN 0-262-13280-X
    1. Insurance, Pollution liability—United States. 2. Insurance, Pollution liability. 3. Air—Pollution—Economic aspects—United States. 4. Air—Pollution—Economic aspects. 5. Carbon Dioxide. I. Richels, Richard G. II. Title.
HG9994.3.M36   1992
368.5—dc20   91-30865
                                                   CIP

For Clare, Jackie, and our offspring, who will live with the consequences of this generation's decisions.

Contents

Preface

This project, initiated in the fall of 1988, was motivated by the events of the preceding summer. A series of congressional hearings had coincided with one of the hottest and driest years on record. With caricatures of an overheated planet appearing on the covers of popular weekly magazines, the greenhouse effect was rapidly evolving from a purely scientific issue into a major public policy debate. Proposals for drastic cuts in emissions were being introduced in both the U.S. Congress and at a number of highly visible international conferences.

As is often the case with complex environmental issues, the analytical base has lagged behind the policy process. Sensible public policy requires balancing benefits and costs. Before committing to a path that would cause a major restructuring of the world's energy system, two questions need to be addressed: (1) What will reductions in emissions buy in terms of reduced environmental damages? (2) What will be the price tag? There is no straightforward answer to either question. Huge gaps remain in our understanding of the physical and biological processes that make up the climate system. Increased concentrations of greenhouse gases could lead to global warming. But by how much? Over what time frame? And what will be the impacts on different regions of the earth?

There are two broad schools of thought regarding the policy implications of these scientific uncertainties. The proponents of immediate controls acknowledge the uncertainties but contend that emissions abatement can be justified solely from an insurance perspective. They argue that low-cost alternatives to carbon-intensive fuels are readily available. Given the stakes, it would be reckless to wait for greater scientific consensus. All that is needed is the political will to engineer the transition to a low-carbon economy. If it is true that emissions can be reduced significantly at little or no cost, emission constraints make a good deal of sense. Immediate controls represent a reasonable hedge against unacceptably rapid climate change. The second school of thought consists of those who are less sanguine about the costs of emissions abatement. If economically attractive alternatives are currently in existence, what is preventing them from automatically entering the marketplace? Fossil fuels provide more than 90 percent of the world's commercial energy. Before obtaining a better understanding of what is at stake, it would be reckless to incur the costs entailed by a rapid transition away from carbon-intensive fuels.

These differing opinions present a dilemma for decision makers. Depending on one's views of control costs, a case can be made either for or against emission cuts. The issue is similar to purchasing an insurance policy. If one believes that there are great risks from global warming and that the insurance premium is negligible, there is little reason to delay. This is the attractiveness of "no regrets" strategies, such as costless conservation. The problem becomes more complex when there are price tags attached to limiting the emission of greenhouse gases. If the insurance premium is expensive, it may be worthwhile to pursue alternatives to immediate cutbacks on emissions.

This book focuses on the costs of reducing carbon dioxide emissions. CO_2 is believed to be the single largest contributor to the problem of global warming and is commanding the most attention within the international community. We examine the costs of emissions abatement from the perspective of both the technology optimist and the pessimist, and we explore the implications for policymaking.

In addition to limiting CO_2 emissions, we examine other forms of greenhouse insurance. Among the options are continuing research to reduce the uncertainties related to climate change and to develop new supply and conservation technologies. Policy makers must decide how to divide the greenhouse insurance dollar among competing needs. What portion goes to the immediate abatement of emissions? What portion goes to resolving scientific uncertainties? And what portion goes to technology development?

Although we have focused on the United States, we have also tried to take a global perspective. We calculate carbon emissions for each of five geopolitical regions under an unconstrained business-as-usual future. We explore possible ways of defining a global CO_2 agreement, compare the impacts at the regional level and estimate the size of the carbon tax required to induce consumers to reduce their dependence on carbon-based fuels. Throughout, we have concentrated on estimating the costs of CO_2 emission abatement and have not attempted to estimate the benefits of slowing the rate of climate change.

At the outset, we had not expected to write a book. Instead we planned to publish a series of separate papers reporting on various aspects of our work. Somewhere along the way, these reports began to take on the appearance of individual chapters, and it seemed logical to combine them into a book. Although this is the first time that our work has been published using the set of assumptions presented here, earlier versions of several of the chapters have appeared elsewhere:

"Emission Limits: An Economic Cost Analysis for the USA," *Energy Journal* 11, no. 2, April 1990.

"The Costs of Reducing U.S. CO_2 Emissions—Further Sensitivity Analyses," *Energy Journal* 11, no. 4, October 1990.

"Global CO_2 Emission Reductions—the Impacts of Rising Energy Costs," *Energy Journal* 12, no. 1, January 1991.

"International Trade in Carbon Emission Rights: A Decomposition Procedure," *American Economic Association Papers and Proceedings* 81, no. 2, May 1991.

"Global 2100: An Almost Consistent Model of CO_2 Emission Limits," *Swiss Journal of Economics and Statistics* 127, no. 2, 1991.

"Buying Greenhouse Insurance," *Energy Policy* 19, no. 6, July–August 1991.

We have received advice and encouragement from many people. We are particularly indebted to Stephen Peck, who initially suggested that we examine the greenhouse issue and has subsequently provided a number of insightful comments. William Hogan has made helpful suggestions throughout the course of this project. We have benefited from the pioneering work on global CO_2 analysis by Jae Edmonds, William Nordhaus, John Reilly, and Gary Yohe and have had profitable discussions with each of them.

Drafts of earlier chapters were reviewed by George Booras, Mark Chaitkin, Hung-po Chao, William Cline, Gregory DeCroix, Robert Dorfman, Michael Gluckman, Lawrence Goulder, Michael Grubb, James Hammitt, George Hidy, Hillard Huntington, Dale Jorgenson, Peter Laut, Lester Lave, Henry Lee, Leonard Levin, Lu Yingzhong,

David Montgomery, Ronald Promboin, Scott Rogers, John Rowse, Thomas Rutherford, Lee Schipper, Leo Schrattenholzer, Daniel Segre , Chauncey Starr, John Stone, James Sweeney, Gary Vine, John Weyant, Robert Williams, and David Wood. We are most grateful to them for commenting on our material in a helpful and constructive manner.

We are much indebted to Stanley Vejtasa, who helped assemble the technology database that underlies our energy supply estimates. It was a pleasure to have had Diane Erdman, Lawrence Gallant, and Robert Luenberger as our research assistants. Lynda Clark, Tola Minkoff, and Daphna Rubin provided invaluable assistance in the production of this book.

Our deepest thanks and appreciation go to our wives, Jackie Manne and Clare Richels. They were often left to deal with the problems of the twentieth century while their spouses were speculating on those of the twenty-first. We are grateful for their patience and good cheer.

Finally, we are indebted to the Electric Power Research Institute (EPRI) for its financial support. The views presented here are solely those of the individual authors and do not necessarily represent the views of EPRI or its members.

1   Overview

The Greenhouse Debate

The greenhouse debate is short on facts and long on rhetoric. Both the activists and the skeptics play important roles in this debate. The activists—many with backgrounds in the physical sciences— point to the potential for disastrous long-term trends in the global climate. They advocate immediate action to offset the increasing accumulation of greenhouse gases in the earth's atmosphere. They argue that the costs of action are low and the potential benefits high. The skeptics—often with backgrounds in economics—counter by listing the uncertainties in climate projections. They note that limiting the emissions of greenhouse gases could be expensive. Today's consumers are being asked to change their life-styles in order to confer uncertain benefits on future generations. Typically, the skeptics recommend a wait-and-see policy, often accompanied by proposals for additional research on the physical and economic effects of greenhouse gas accumulation.

The greenhouse effect poses a serious dilemma for policy makers. The experts are deadlocked on both the likelihood and the timing of the problem. Enormous uncertainties remain in our understanding of the greenhouse effect, its likely consequences, and the possible effectiveness of various countermeasures. These uncertainties will not be resolved for decades.

The stakes are large. Waiting leads to the risk of irreversible damages. Immediate action leads to the risk that large costs will be incurred in the near future, and there is considerable disagreement on how these actions would eventually affect the world's climate. Policy makers would do well to act as though they were purchasers of greenhouse insurance. They must weigh the possible costs of delay against those of premature action, but also recognize that there are not enough resources available to purchase insurance against all conceivable downside risks.

Uncertainty is but one factor that confounds the debate. The greenhouse problem is a global issue. Many countries are taking the position that if significant efforts are required to reduce emissions, these efforts should be undertaken only in the context of an international agreement. As the climate debate moves toward the consideration of specific legislative initiatives and policy options, international negotiations will become increasingly important.

The negotiation of a greenhouse gas agreement would be extraordinarily complex. Major reductions in emissions could be expensive. Some nations might incur high costs in order to achieve modest reductions, and the converse might hold for others. Each country's leaders need to weigh the benefits and the costs of proposed actions in order to arrive at an overall judgment.

The Costs of Limiting CO_2 Emissions

It is believed that carbon dioxide (CO_2) emissions are currently responsible for more than 50 percent of the human contributions to the greenhouse effect. Accordingly, the energy sector plays an important role in strategies to delay climate change. Over the next few decades, most strategies call for a push toward greater energy efficiency and—to whatever extent is feasible—moving away from coal and oil toward natural gas with its lower carbon emissions per unit of energy. Over the longer term, energy could be supplied by carbon-free alternatives such as solar (in several different forms), nuclear energy, and fusion.

Estimates of the costs of a CO_2 limit vary widely. For a given abatement target, the exact amount will depend on the severity of the carbon limit, the rate of energy conservation, and the supply technologies available for meeting energy demands. Differences in abatement cost estimates arise primarily from alternative views about the potential for innovations in the energy sector. Technology optimists describe an energy future with abundant low-cost, carbon-free supply alternatives and low overall demands for energy. In such a world, carbon-free substitutes would be economically attractive. Highly efficient end-use technologies would virtually eliminate any growth in fossil fuel consumption. Technology pessimists visualize a very different energy future. Coal would be used as the principal source of electric power and of liquid fuels for transportation. The lack of carbon-free substitutes, combined with rapidly rising energy demands, would make it difficult for consumers to reduce their dependence on carbon-intensive fuels.

This book outlines a way to think about greenhouse decisions under uncertainty. It provides region-by-region estimates of the costs that would underlie an international agreement. We focus on just one aspect of the greenhouse debate: the costs of limiting the carbon dioxide emissions produced by burning fossil fuels: coal, oil, and gas. Our work is based on a computer model known as Global 2100. The name emphasizes the global nature of the problem and the need for a long-term perspective.

We analyze the economic impacts of limiting CO_2 emissions under alternative supply and conservation scenarios. Global 2100 is employed to indicate how emissions are likely to evolve in the absence of carbon limits and how regional patterns might shift during the next century. The model provides a consistent way to examine alternative strategies for limiting global emissions and to calculate the impact of higher energy prices on gross domestic product (GDP). It enables us to estimate the size of the tax required to induce consumers to reduce carbon emissions. We also analyze the possibility of significant inter-regional differences in carbon taxes that would lead to opportunities for international trade in emission rights.

The costs of abatement are only part of the story, but they are an essential part. They enable us to assess the feasibility of alternative proposals and to determine which measures are cost-effective. Moreover, a reduction in emissions is not the sole policy response that is available. There is a point at which further reductions could become so expensive that it would be preferable to shift to other options, such as adaptation to climate change. Without careful analysis, it is difficult to know where that point might be.

We do not attempt to estimate the benefits of slowing the rate of climate change through a reduction in worldwide CO_2 emissions. Our analysis is confined to the direct impacts of carbon limits upon the cost of energy and the resulting effects on the economy as a whole. It is a far more formidable task to estimate the benefits from reducing emissions, and is well beyond the scope of this book. Clearly, policy makers will need information on both costs and benefits in order to make a balanced decision. (For imaginative approaches to integrated benefit-cost analysis, see Nordhaus 1991 and Peck and Teisberg 1991.)

The Greenhouse Effect: Likely Causes and Possible Consequences

Before turning to the central focus of this book—the costs of CO_2 emissions abatement—we summarize some general background on the greenhouse effect, its likely causes, and possible consequences. There is a broad consensus on many of these issues, but disagreement on others. In order to follow the debate over the role of carbon dioxide and the energy sector, it is essential to understand the points of contention.

The earth's climate is determined by the balance between energy received from the sun and energy radiated back into space. Slightly more than half of the solar energy entering the atmosphere is absorbed by clouds and particles in the air or is reflected back into space. The remainder is absorbed at the earth's surface and then radiated outward in the form of heat. Rather than escaping directly into space, some of this heat is trapped by traces of atmospheric water vapor, carbon dioxide, and other infrared absorbing gases and re-emitted back to earth. This is the phenomenon termed the greenhouse effect.

The naturally occurring greenhouse effect has warmed the earth for billions of years, and it is essential to life on our planet. Without greenhouse gases, the average temperature of the earth would be about 34° C colder, well below freezing. The extremely cold temperatures on the surface of Mars and the oven-hot surface of Venus can be explained primarily by differing atmospheric levels of CO_2. By contrast, the composition of the earth's atmosphere is ideal for supporting life.

Climate scientists are concerned that human activities are increasing the atmospheric concentrations of the naturally occurring greenhouse gases and that we are compounding the difficulties by adding potent new gases such as the chlorofluorocarbons (CFCs). If recent trends continue, the buildup of these gases will enhance the greenhouse effect and could cause significant warming within the next century. Enormous uncertainty surrounds virtually every aspect of climate change. How much warming will occur? How quickly? What will be the region-by-region consequences?

Figure 1.1 and table 1.1 summarize what is known about the key greenhouse gases. Ice core studies show that there have been major increases in these gases since the Industrial Revolution. Many observers believe that human activities account for much of the 0.6° C

Table 1.1
Summary of Key Greenhouse Gases Influenced by Human Activities

Parameter	CO_2	CH_4	CFC-11	CFC-12	N_2O
Preindustrial atmospheric concentration (1750–1800)	280 ppmv[a]	0.8 ppmv	0	0	288 ppbv[a]
Current atmospheric concentration (1990)[b]	353 ppmv	1.72 ppmv	280 pptv[a]	484 pptv	310 ppbv
Current rate of annual atmospheric accumulation	1.8 ppmv (0.5%)	0.015 ppmv (0.9%)	9.5 pptv (4%)	17 pptv (4%)	0.8 pptv (0.25%)
Atmospheric lifetime[c] (years)	(50–200)	10	65	130	150

Source: Houghton et al. (1990).

Note: Ozone has not been included in the table because of lack of precise data.

a. ppmv-parts per million by volume; ppbv-parts per billion by volume; pptv-parts per trillion by volume.

b. The current concentrations have been estimated based on an extrapolation of measurements reported for earlier years, assuming that the recent trends remained approximately constant.

c. For each gas in the table except CO_2, "lifetime" is defined as the ratio of the atmospheric content to the total rate of removal. This time scale also characterizes the rate of adjustment of the atmospheric concentrations if the emission rates are changed abruptly. CO_2 is a special case since it has no real sinks, but is merely circulated between various reservoirs (atmosphere, ocean, biota). The lifetime of CO_2 given in the table is a rough indication of the time it would take for the CO_2 concentration to adjust to changes in the emissions.

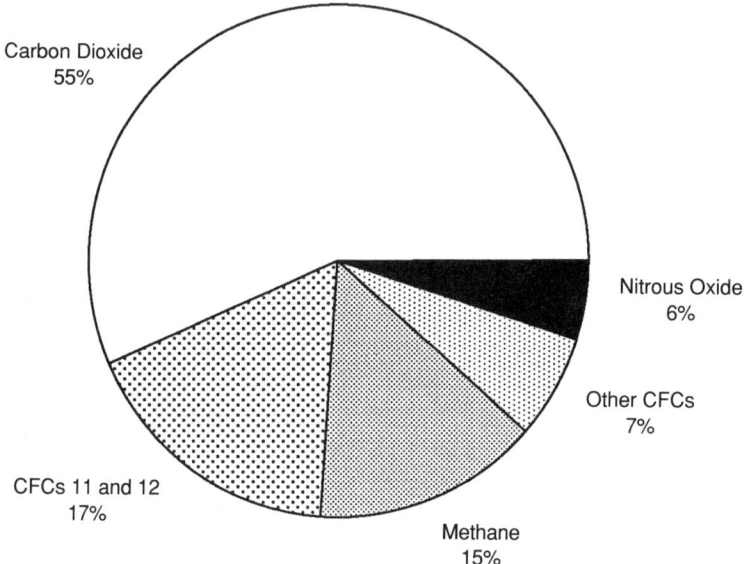

Figure 1.1 Human contributions to the greenhouse effect. Source: Houghton et al. (1990).

rise in average global temperature observed over the last century. Although the recorded temperature increase is consistent with the predicted effects of CO_2 and other greenhouse gases, there is a plausible alternative explanation: chance variations within the normal range of temperature trends.

Atmospheric scientists rely on computer models to predict how key climate variables (temperature, rainfall, wind speed, humidity) might change as a result of increases in the major greenhouse gases. The atmospheric models, originally developed for long-term weather forecasting, have a number of limitations (World Resources Institute 1990). The shortcomings include poor spatial resolution, inadequate accounting for various feedback mechanisms that could exacerbate or counteract the greenhouse effect, and insufficient treatment of such important factors as variations in solar output, volcanic activity, and the earth's reflectivity.

These limitations lead to inaccuracies in the projections of mean global temperature. Nevertheless, the best evidence suggests that a doubling of greenhouse gas concentrations from their pre–Industrial Revolution levels would increase average global temperatures from 1.5 to 4.5° C, and that the doubling of concentrations could take place well within the next century (National Research Council 1983.) To provide some historical perspective, a 1.5° C rise would produce the warmest temperatures seen on earth in 6,000 years. A 4.5° C increase would raise the world's temperature to a level that was last experienced in the age of dinosaurs. The earth could be a very different planet from the one that we know today.

This leads to the question of impacts on future generations. Although a growing amount of research is being done in this area, information is still at a rudimentary stage (Department of Energy Multi-Laboratory Climate Change Committee 1990). Today's climate models can often replicate events at continental scales, but not for individual regions. Events that are essential for impact analysis (rainfall, storms, and floods) occur at a local and regional level. Until we are able to do a better job at regional forecasting, the assessment of the environmental, economic, and social consequences of global warming will remain an elusive goal.

The risks may be substantial. Climate change has the potential to alter the natural setting of the earth. A temperature increase of several degrees could lead to severe droughts in the grain belt and other productive agriculture regions. The health and productivity of many forest areas may be reduced. The sea level may rise, inundating wetlands and coastal lowlands and accelerating coastal erosion. Currently threatened and endangered species could face greater risks of extinction.

Although none of these impacts is certain, the possibility offers good reasons for proceeding cautiously. These concerns have resulted in a number of international proposals to reduce greenhouse emissions. A CFC reduction treaty has already been signed, and carbon dioxide seems to be next in the order of priority for greenhouse gas negotiations. The climate models disagree on the absolute temperature rise, but they are fairly consistent in estimating that carbon dioxide could account for more than half the projected cumulative temperature increase. The remainder is allocated across the other greenhouse gases. This explains the emphasis on CO_2 in the negotiations related to climate change.

Carbon Dioxide, the Energy Sector and Economic Growth

Two principal types of human activities affect the flow of carbon dioxide: the combustion of fossil fuels and changes in land use practices

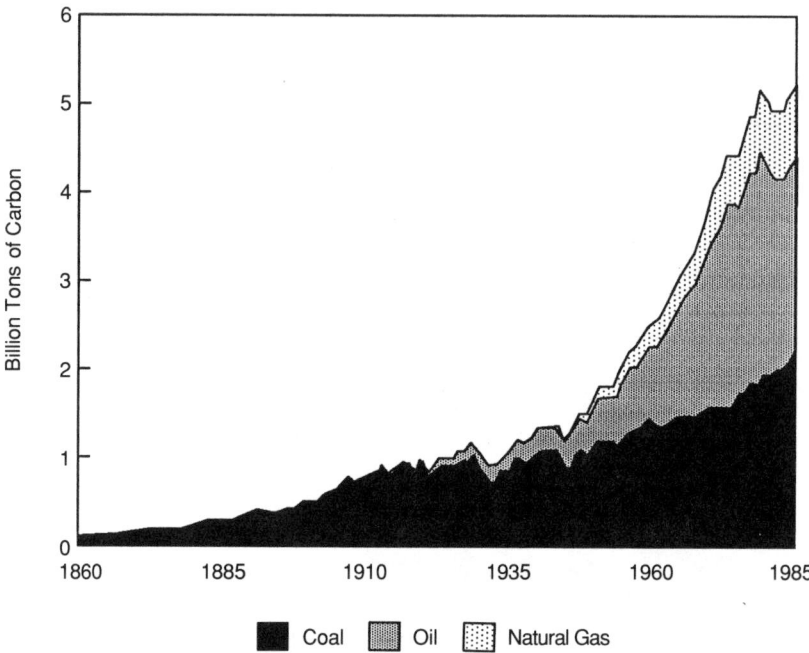

Figure 1.2 Carbon emissions due to fossil fuel consumption 1860–1985. Source: Boden et al. (1990).

(deforestation, abandonment of farmland). For the world as a whole, the burning of fossil fuels produced about 6 billion tons of carbon emissions in 1990. It is much more difficult to measure the contribution from changing land use practices. The best available estimates suggest a net release of about 2 billion tons annually, but some observers estimate that the contribution may be closer to zero or possibly negative (Krause et al. 1990).

Although CO_2 is emitted whenever fossil fuels are burned, the rate of emissions varies significantly among fuels. Coal's carbon emission coefficient is highest. Relative to oil, it produces 21 percent more CO_2 per unit of energy consumption. Relative to natural gas, it produces 76 percent more CO_2. Carbon-free fuels (hydroelectricity, nuclear, and solar) may have other environmental impacts, but they emit no CO_2.

Figure 1.2 summarizes the historical record of global emissions from fossil fuel consumption. Between 1860 and 1985, emissions increased sixtyfold. Over time, the industrialized countries have economized on labor and capital by shifting their fuel mix away from coal and toward oil, gas, and carbon-free fuels. As a by-product of this fuel switching, they automatically lowered their carbon emissions per unit of energy consumption. Conversely, there may come a time when low-cost oil and gas resources are exhausted, and it becomes economical to produce synthetic fuels from coal or shale oil. This could lead to a significant increase in the carbon emissions required to produce a given amount of liquid fuels for transportation or other uses.

Currently, the industrialized countries are responsible for more than 70 percent of the CO_2 resulting from the combustion of fossil fuels. Figure 1.3 ranks the six countries that emitted the greatest amounts in 1988. It also reports emissions on a per capita basis and shows how emissions in the industrialized countries greatly exceed those per person in China and India. These statistics refer to a single point in time. Energy consumption patterns have been shifting and are likely to continue to shift. History may provide some guidance as to where we might be heading.

Figure 1.4 shows global emissions between 1950 and 1985 for industrialized and developing countries, and it highlights two important trends: a slowing in the overall growth rate following the 1973 oil price explosion and a reduction in the share of the industrialized relative to the developing countries. These trends also emerge from a comparison of individual country profiles. Figure 1.5 contains the growth rates of emissions for the six largest contributors and for the world. The post–World War II era is divided into two periods: 1950–1970 and 1971–1988. During the later years, the three Organisation for Economic Co-operation and Development (OECD) countries (United States, Japan, and West Germany) grew at rates well below the global average. By contrast, China and India grew three times as rapidly as the global average. In part, the slowdown in the OECD countries may be explained by lower rates of economic growth than in the developing countries. Also at work is a gradual decoupling between energy and GDP growth—and a decoupling between CO_2 and energy use. For example, in the United States, the GDP and CO_2 emissions grew at annual rates of 3.2 percent and 1.9 percent, respectively, between 1950 and 1985.

The factors underlying the decoupling are based upon the following identity (Kaya 1989):

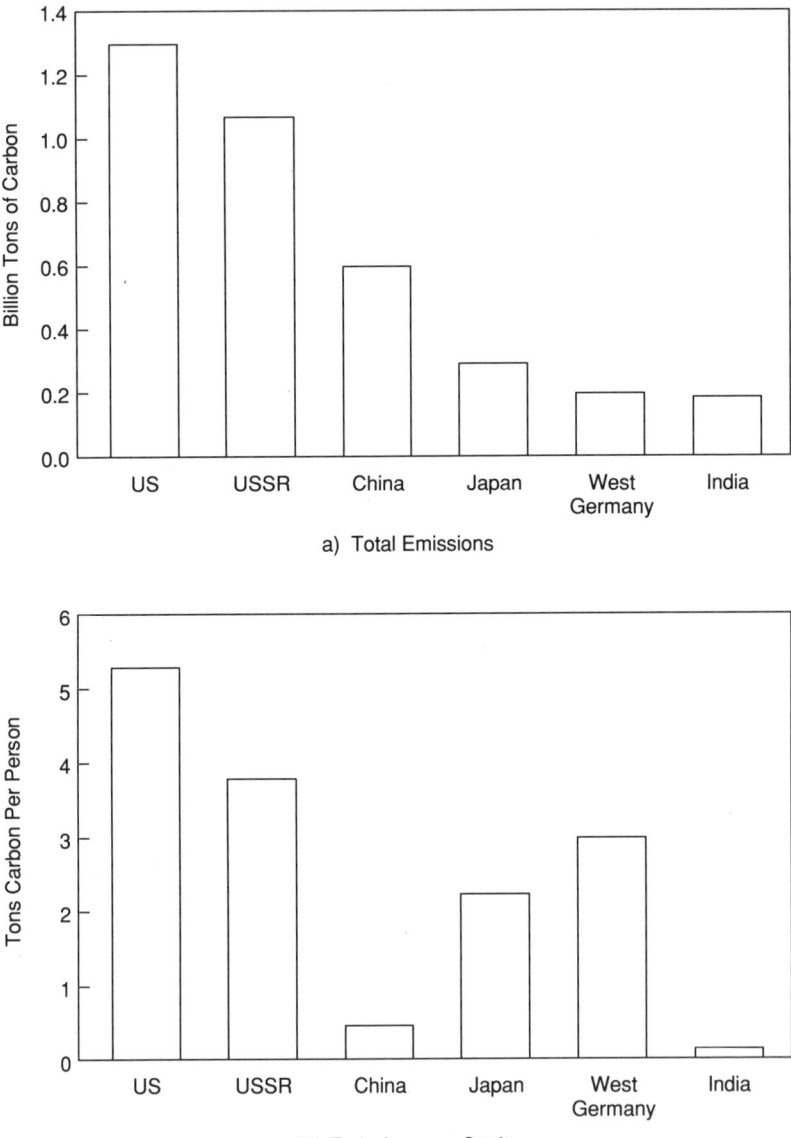

Figure 1.3 Carbon emissions in 1988 for selected countries. Source: Boden et al. (1990).

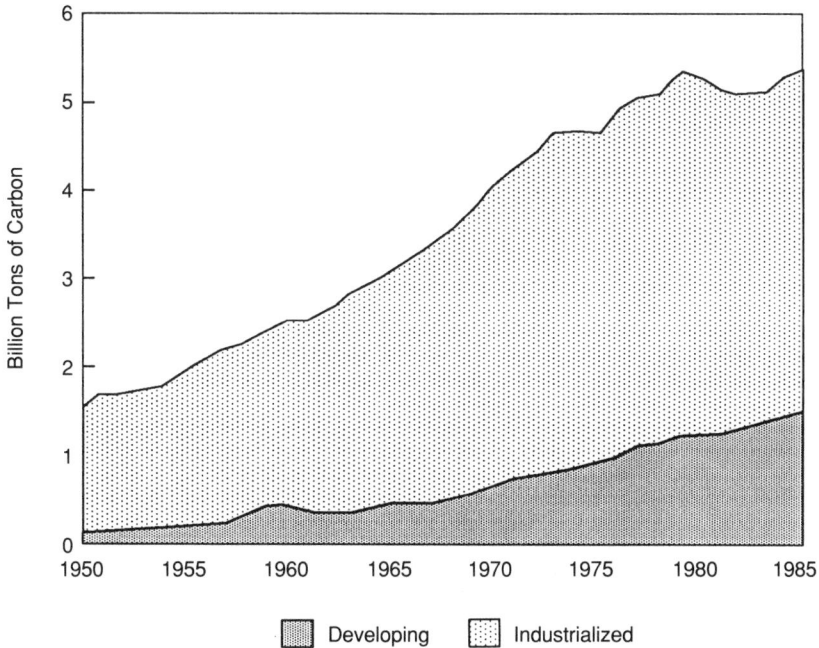

Figure 1.4 Carbon emissions growth rates for industrialized and developing countries 1950–1985. Source: Krause et al. (1990).

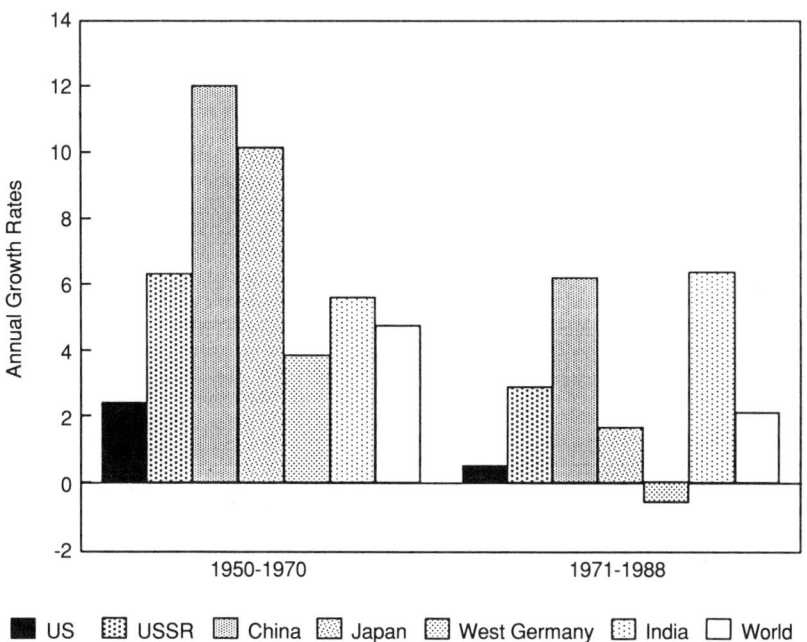

Figure 1.5 Carbon emissions growth rates for selected countries and the world

Growth rate of output	3.2%
Decline rate of energy use per unit of output	−0.8
Decline rate of CO_2 emissions per unit of energy use	−0.5
Growth rate of CO_2 emissions	1.9%

The emissions level is determined not only by the economic growth rate but also by changes in energy consumption per unit of output—an average annual decline of 0.8 percent during this period. In part, this was the result of efficiency gains induced by the oil price shocks of 1973 and 1979. Technological progress and structural economic changes have also reduced the amount of energy required per unit of GDP.

The third term in Kaya's identity—the reduction in CO_2 per unit of energy use—is related to fuel switching within the energy sector. During this period, there was a shift away from coal—hence, an annual decline of 0.5 percent in the growth rate of CO_2 emissions per unit of energy use. In 1950, coal was responsible for over 37 percent of the United States primary energy consumption. By 1985, its share had dropped to less than 25 percent.

This identity is also useful in explaining the high emissions growth rates for China and India. Emissions increased at annual rates of 6 percent in both countries during the period 1971 to 1988. In China, the average annual GDP growth rate has been reported as 9 percent, but there have been significant reductions in energy use per unit of output. Energy intensity has declined at about 3 percent per year. For India, the story is quite different. CO_2 growth rates have actually outpaced GDP (6 percent versus 4.2 percent). Structural changes have led to a greater share of energy-intensive industries in total output—hence a rise in energy intensity for the Indian economy as a whole.

Future Patterns of Growth

What does the historical record tell us about the future? Clearly, population and per capita economic growth will continue to be major determinants of CO_2 emissions. If global income inequalities are to be diminished, developing countries must experience relatively high rates of economic growth. Figure 1.6 shows our business-as-usual global projections for the twenty-first century. These are represented as index numbers, with 1990 = 100. It is expected that the world's population will double during the first half of the twenty-first century,

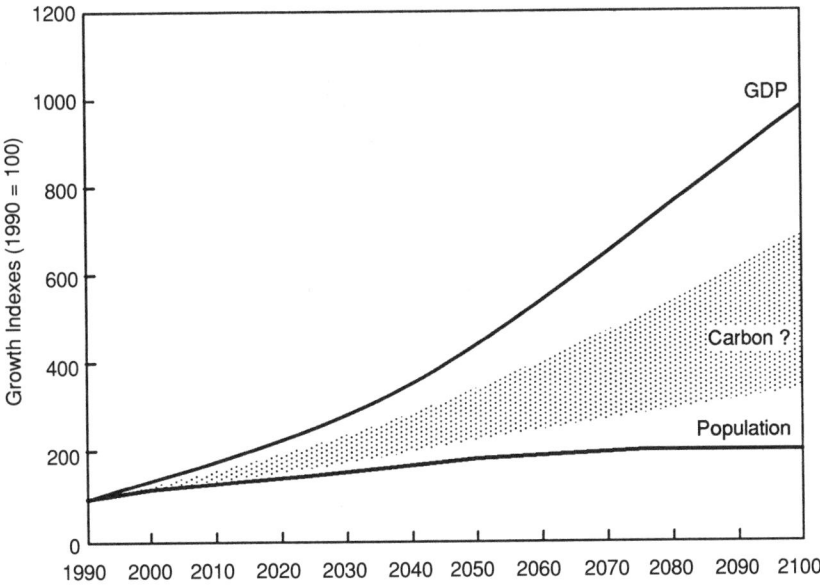

Figure 1.6   Global growth indexes, business as usual

but will stop growing thereafter. Virtually all of the increase will take place in the developing countries, but eventually their population will stabilize, just as has already happened in most of the industrialized countries. (See Zachariah and Vu 1988.)

Our GDP projections represent the average of the Intergovernmental Panel on Climate Change's (IPCC) high- and low-growth scenarios (Working Group III 1990). According to figure 1.6, global GDP is likely to grow nearly tenfold during the twenty-first century and will far outstrip the rate of population increase. This GDP growth seems enormous, but it follows directly from the extrapolation of modest improvements in per capita productivity. That is, if productivity expands at 1.5 percent annually for 110 years, this would compound to more than a fivefold increase in per capita output. When we allow for a doubling of the world's population, this leads to a tenfold increase in GDP.

Carbon emissions are likely to grow but not as rapidly as the GDP. Over time they may be decoupled. Fuel switching and reductions in energy intensity have played important roles in the past and are likely to remain important. Future emissions will depend on the availability

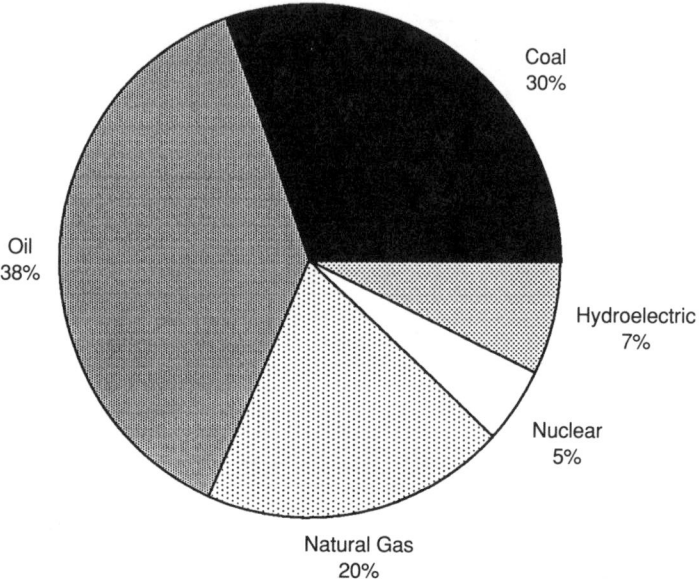

Figure 1.7 World energy consumption, 1988. Note: Excludes biomass.

of economically attractive carbon-free supply alternatives and cost-effective options for reducing energy demands. Considering the enormous uncertainty surrounding each of these factors, it is not surprising that there is so much disagreement over the meaning of a business-as-usual future.

Prospects for Fuel Switching

Figure 1.7 shows 1988 global energy consumption, excluding that from biomass. There are good reasons for the world's heavy dependence on fossil fuels. They are convenient to use, and the total resource base is enormous. Among the fossil fuels, there are important differences in availability and carbon content. Coal is the most abundant, but it also emits the most carbon per unit of useful energy.

At the current worldwide rate of consumption, there is enough coal to last for more than 1,000 years. By contrast, the supplies of conventional oil and gas are likely to be depleted in less than a century.

From the perspective of resource availability, coal constitutes a logical bridge to the future. If no CO_2 limits are imposed, coal could resume its role as the dominant fossil fuel in both the electric and nonelectric sector. In many countries, gas-fired electric generating units have a competitive advantage over coal at 1990 prices. As natural gas resources become exhausted, the price of gas is likely to rise, and coal would replace gas as the principal fuel for electricity generation during the next few decades.

In the nonelectric sector, it will be difficult for coal to replace oil directly as a transportation fuel. With the eventual exhaustion of conventional oil supplies and a rise in crude oil prices, coal-based synthetics could become a viable alternative for meeting nonelectric energy needs. Unfortunately, the coal-based substitutes produce twice as much CO_2 per unit of useful energy as oil. This would mean a substantial growth in carbon emissions under a business-as-usual energy future.

If carbon limits are imposed, coal's role would be limited. With carbon constraints, high penalties would be imposed on carbon-intensive fuels. Consumers would have to be induced to switch to lower-carbon alternatives. In the near term, the mix of energy supplies could be shifted away from coal and oil toward natural gas. The costs of such a strategy are likely to differ among regions. Figure 1.8 shows our estimates of gas supplies for various parts of the world. These resource endowments differ widely. Fulkerson et al. (1990) estimate that if natural gas replaced coal in all uses, United States supplies would last for only eighteen years. By contrast, the Soviet Union's supplies would last for seventy to eighty years. Gas can be traded via pipeline and LNG (liquefied natural gas) shipments, but China, India, and other countries with abundant supplies of coal may be unable to afford an increase in their dependence on external sources of gas supply. Figure 1.8 also shows estimates of conventional oil supplies by region. CO_2 limitations would affect the international oil market in different ways over time. In the near term, a carbon constraint would depress oil demands and impose downward pressures on international oil prices. Over the longer term, these limitations would inhibit the use of coal-based synthetic fuels and could lead to a premium value for crude oil.

Eventually a carbon constraint would lead us to base our energy supplies on carbon-free sources, such as solar, fission, and fusion. There is wide disagreement over the prospects for such technologies.

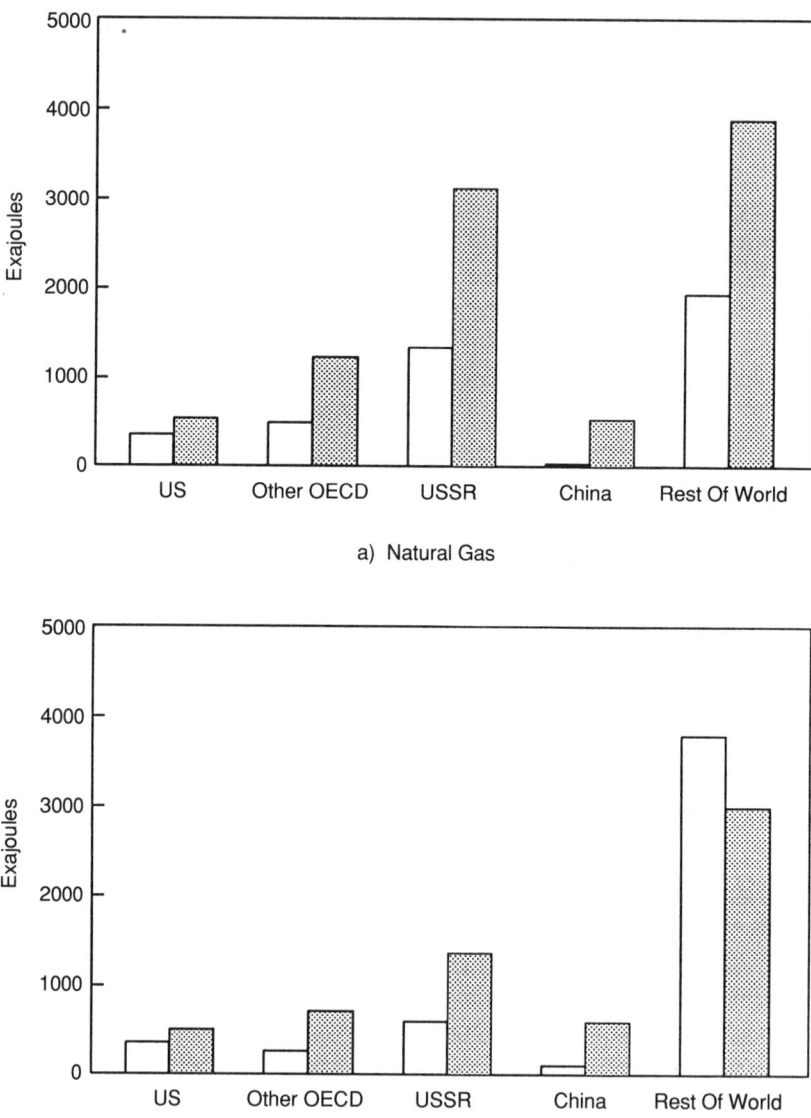

a) Natural Gas

b) Oil

Reserves ☐   Resources ▦

Figure 1.8 Economically recoverable natural gas and oil supplies

When evaluating advanced systems, consideration must be given to technical feasibility; capital, fuel, and resource costs; health, safety, and environmental impacts; and the time scale required for large-scale development. It is difficult to predict the ultimate potential and the inherent limitations of technologies that are still on the drawing board. At this point, we must be satisfied with placing bounds on the range of uncertainty. This can be done by posing a series of what-if questions designed to highlight the divergence of viewpoints.

The "Efficiency Gap"

Few other issues have engendered as much controversy as the "efficiency gap"—the proposition that there is a large gap between the energy conservation practices that are economically feasible and those currently employed. Through "technology forcing," we could simultaneously reduce energy consumption, carbon emissions, and the lifecycle costs borne by energy consumers.

In assessing the costs of carbon limits, there are several schools of thought. At one end of the spectrum are those who believe that the direct costs of reducing carbon emissions would be negligible and perhaps even negative. They argue that market failures limit investment in cost-effective conservation and in renewable forms of energy; there would be little or no cost to removing the barriers to a more efficient use of energy; and carbon limits will lead consumers and producers to adopt technologies that reduce carbon emissions and also reduce energy utilization costs (Lovins and Lovins 1990). Toward the other end of the spectrum are those who believe that energy markets are already operating efficiently. If untapped efficiency improvements would reduce costs for individual consumers and producers, why are these improvements not taking place automatically? This school holds that the costs of mandatory efficiency standards are often hidden. When one takes account of the full range of consumer preferences, there is no free lunch. (For more on this perspective, see Council of Economic Advisers 1990.)

The controversy stems in part from the enormous intercountry variability of energy consumption per unit of output. International differences are often cited as proof of sizable slack in some energy systems, particularly in those of the United States, the Soviet Union, China, and Canada. No doubt, there are cost-effective opportunities for improved energy efficiency in each of these countries. However, one must be careful when making international comparisons. A substantial amount of the observed differences can be explained by the land mass, population density, and composition of economic activity in these countries.

Historical differences must also be taken into account. The United States had access to abundant supplies of low-cost energy at virtually every stage of its industrialization process. As a result, it tended to substitute energy in place of capital and labor wherever possible. Rosenberg (1989) notes that during the development of the American West, homes were heated by larger and less efficient wood-burning fireplaces than those used in Europe. Although the fireplaces generated less useful heat per log, the large size of the logs significantly reduced the labor requirements for wood chopping.

This example points to a consideration often overlooked in the debate over energy efficiency: households and firms are concerned with minimizing their total costs—not just the costs of energy per unit of output. It may make sense to substitute capital, labor, and other inputs in place of energy but only if it is economically efficient to do so. Intercountry differences in energy per unit of GDP can be explained partly by differences in relative prices. Widespread intercountry differences in energy use do not in themselves reflect irrational choices.

We do not mean to imply that energy markets are behaving perfectly and that all options for costless efficiency improvements are being adopted. To the contrary, essentially all analysts agree that there are cost-effective opportunities for improving energy efficiency and for reducing CO_2 emissions. The disagreement is over the size and timing of these opportunities.

Why would a country like the United States invest too little in energy efficiency? How could we fail to consume a free lunch? Often-cited barriers to energy efficiency include the following: consumers confront difficulties when they seek information on the options for improving energy efficiency; consumers act as though they have abnormally high discount rates when making appliance purchase decisions; and electricity prices have tended to be based on average rather than marginal costs of supply.

There is also the problem of externalities. Retail fuel prices do not necessarily reflect the full costs associated with the production, transportation, and conversion of fuels. Inadequate charges may be levied for externalities such as air and water pollution, the storage and disposal of nuclear wastes, and increasing dependence on oil imports from the Middle East.

The efficiency gap controversy centers on the costs and political feasibility of removing these types of market imperfections. The resulting gains are not always justifiable. For example, proponents claim that an increase in the U.S. CAFE (corporate average fuel economy) mandatory automobile standards will reduce gasoline consumption at little or no cost. CAFE opponents counter by pointing to the reductions in safety and comfort that would occur as manufacturers reduced automobile size in order to meet higher fuel efficiency standards. Moreover, an improvement in fuel efficiency lowers the cost per kilometer driven and has the perverse effect of encouraging more trips. In view of these considerations, the U.S. Council of Economic Advisers (1990) has expressed doubts over the potential energy savings from mandatory auto efficiency standards.

There are similar disagreements over energy savings in the electric sector. One survey notes that the estimates of potential savings range from 4 to 75 percent of energy consumption (Interagency Task Force 1990). The broad range reflects alternative views about the costs of correcting market failures, as well as the net savings that are achievable in practice.

Over the long term, the price mechanism may prove to be more important than the correction of market failures. As conventional oil and gas resources become depleted and energy becomes increasingly expensive, it will be attractive to adopt energy-efficient end-use technologies. This would offer an opportunity to reduce both total costs and the energy costs per unit of output. Such adjustments will take time. Energy-efficient systems are often embeddded in long-lived durable goods (autos, housing, equipment, structures), and these will not be replaced overnight.

Global 2100 provides a means of examining a broad range of views on the size of the efficiency gap. We do not assume that energy markets are behaving perfectly and that every option is being adopted for costless improvements in energy efficiency. We allow for the possibility of both price-induced and nonprice conservation. This book will not resolve the issue of the efficiency gap. Since 1973, the world has been awash with suggestions for improving energy efficiency. Some have worked, others have failed. It remains to be proved whether there are huge gains to be had at little or no cost. As with the supply-side uncertainties, we can pose a series of what-if questions, highlight the divergence of viewpoints, and assess their impact on the overall costs of a carbon constraint.

Outline of This Book

The remainder of this book is divided into two parts. Chapters 2 through 6 are addressed to the public at large and to our representatives at the international CO_2 negotiating table. Chapter 2 provides an overview of Global 2100. It contains a description of the supply and demand sectors employed within the analysis, the rationale for the model's design, and a discussion of the macroeconomic parameters that affect the demand for energy.

Chapter 3 focuses on the United States. First we examine how the energy sector is likely to evolve under business-as-usual conditions. We then explore the costs of alternative carbon constraints. The impacts of a CO_2 limit will depend on the technologies and the resources for meeting energy demands, as well as the demands themselves. Overall costs are explored from alternative perspectives encompassing a broad range of views about the future character of the energy sector.

In chapter 4, we look at the issue of greenhouse insurance. Given the deadlock among climate experts on the chances of a global calamity, what steps should we take today so as to reduce the risks to future generations? Three forms of insurance are analyzed: continued intensive science research to reduce climate and impact uncertainties, development of new supply and conservation technologies to reduce abatement costs, and immediate reductions in emissions in order to slow climate changes.

Chapter 5 provides a global perspective. We begin by projecting the size and pattern of emissions in the absence of measures to slow growth. Carbon emissions are calculated for each of five geopolitical regions under an unconstrained business-as-usual future. We explore possible ways of defining a global CO_2 agreement, compare the impacts at the regional level, and estimate the size of the carbon tax required in each region to induce consumers to reduce their dependence on carbon-based fuels. To the extent that there are region-by-region differences in carbon tax rates, it would be useful to establish an international market in emission rights. Accordingly, we provide an estimate of the benefits that might result from trade in these rights.

Chapter 6 shows that much of our analysis can be summarized in terms of willingness-to-pay functions—the value of carbon that is associated with each possible level of emissions curtailment. By ignoring the dynamics of the transition away from conventional oil and gas, we show that many of the qualitative results produced by Global 2100 can be explained through simpler models.

Part II is more technical in nature and is addressed to modeling specialists. Chapter 7 documents the algebraic formulation employed in Global 2100. Chapter 8 describes the decomposition procedure used for analyzing international trade in carbon rights. Finally, Chapter 9 provides a backcasting analysis to estimate the energy conservation parameters for the United States over the years 1960 through 1990. This does not validate our projections for the twenty-first century, but it provides a historical cross-check upon them. Our numerical assumptions are listed in appendixes A through E.

References

T. Boden, P. Kanciruk, and M. Farrell. 1990. *Trends '90*. Carbon Dioxide Information Analysis Center, Oak Ridge, Tenn.

Council of Economic Advisers. 1990. *Economic Report of the President*. U. S. Government Printing Office, Washington, D.C.

Department of Energy Multi-Laboratory Climate Change Committee. 1990. *Energy and Climate Change*. Lewis Publishers, Chelsea, Mich.

B. Fulkerson, R. Judkins, and M. Sanghvi. 1990. "Energy from Fossil Fuels." *Scientific American* 263 (3), September.

J. T. Houghton, G. J. Jenkins, and J. J. Ephraums. 1990. *Climate Change—The IPCC Scientific Assessment*. Cambridge University Press, Cambridge.

Interagency Task Force. 1990. *The Economics of Long-Term Global Climate Change—A Preliminary Assessment*. Office of Policy, Planning and Analysis, U.S. Department of Energy, Washington, D.C.

Y. Kaya. 1989. "Impact of Carbon Dioxide Emission Control on GNP Growth: Interpretation of Proposed Scenarios." Intergovernmental Panel on Climate Change/Response Strategies Working Group. May.

A. Lovins and L. Lovins. 1990. "Least-Cost Climate Stabilization," Rocky Mountain Institute, Old Snowmass, Colo.

F. Krause et al. 1990. *Energy Policy in the Greenhouse*. Earthscan Publications Ltd., London.

National Research Council. 1983. *Changing Climate*. National Academy Press, Washington, D.C.

W. D. Nordhaus. 1991. "Economic Approaches to Greenhouse Warming." In R. Dornbusch and J. M. Poterba (eds.), *Global Warming*. MIT Press, Cambridge, Mass.

S. Peck and T. Teisberg. 1991. "CETA: A Model for Carbon Emissions Trajectory Assessment." Electric Power Research Institute and Teisberg Associates, February.

N. Rosenberg. 1989. "Energy Efficient Technologies: Past And Future Perspectives." Stanford University, Stanford, Calif.

Working Group III. 1990. *Formulation of Response Strategies*. Intergovernmental Panel on Climate Change. Island Press, Washington, D.C.

World Resources Institute. 1990. *World Resources 1990–1991*. Oxford University Press, New York.

K. C. Zachariah and M. T. Vu. 1988. *World Population Projections, 1987–1988 Edition*. World Bank, Johns Hopkins University Press, Baltimore, Md.

I   Policy Perspectives

2

Global 2100: A Model of CO_2-Energy-Economy Interactions

Introduction

Energy costs will be affected by limits on CO_2 emissions. In this book, we consider the likely effects of rising energy prices on both the energy sector and the economy as a whole. To do this, it is necessary to understand the major interrelationships within the energy sector and the connection between energy and economic growth. The Global 2100 model has been constructed especially for this purpose and serves as our basic tool.

In designing Global 2100, we tried to adhere to three criteria: simplicity, transparency, and ease of exploring a wide range of assumptions. It is impractical to construct a single model to address all questions related to energy and environmental policy. Global 2100 provides no more details than are needed to deal with the specifics of the debate over carbon limitations. The model is designed for exploring broad alternatives, not for detailed analysis within the energy sector of a given country or region. For this reason, we do not distinguish among energy types in detail or analyze demands in a disaggregated way. Although this is a top-down approach, it does not automatically lead to qualitatively different conclusions from bottom-up methods. In fact, Global 2100 is designed so that the two approaches can be used in a complementary fashion.

Transparency is important. Results should not be trusted unless they are intuitively understandable. In this type of analysis, the general conclusions typically follow from a few basic facts and assumptions. It is important that critical assumptions be highlighted and that one be able to understand which of them determine the basic results. We describe the fundamental relationships in the model and see how they follow from general considerations about the energy sector and the economy as a whole.

The model is capable of exploring a wide range of assumptions. Energy forecasting is a difficult business. The impacts of today's decisions will unfold over a period of a century or more. Future supplies and demands depend on a combination of highly uncertain market forces, social developments, and public policies. Rather than try to predict what the future will look like, we have designed a tool for addressing a series of what-if questions. Critical quantitative parameters can be varied easily to explore a wide range of assumptions about the future character of the energy system.

Our computer simulations are designed as controlled experiments. Specifically, we distinguish between two debates: one over the efficiency gap and the other over the cost of carbon constraints. For any specific set of supply and demand assumptions, we examine how the economy might perform in the best way possible, with and without the imposition of carbon limits. To the extent that there is an efficiency gap, we take advantage of it in both scenarios.

With this type of controlled experiment, we automatically rule out the possibility that carbon limits will improve the performance of the economy. To demonstrate that we can save carbon and at the same time improve economic performance, we would have to adopt a specific set of assumptions about the behavior of political systems. That is, decision makers would fail to adopt energy conservation measures that are desirable in themselves but would reach a consensus on such measures if they became convinced of a greenhouse threat. (For an example of this type of reasoning, see the comparison between technological costing and energy modeling of CO_2 mitigation options reported in National Academy of Sciences 1991.)

There is no easy way to validate assumptions about political behavior. Global 2100 is designed only to promote second-order agreement: defining the specific areas of disagreement among the participants in the greenhouse debate. The model is not intended to provide definitive numerical forecasts but rather to provide a logical framework for thinking about the trade-offs between costs and emission reductions.

Model Overview

We chose the name Global 2100 in order to emphasize both the global nature of the carbon emissions problem and the need for a long-term perspective. There are long time lags inherent in the buildup of CO$_2$ and in the transition away from carbon-based fuels. Our model is benchmarked against 1990 base year statistics, and the projections cover eleven ten-year time intervals extending from 2000 through 2100.

The globe is divided into five major geopolitical groupings: (1) the United States, (2) other OECD nations (Western Europe, Canada, Japan, Australia, and New Zealand), (3) the former Soviet Union, (4) China, and (5) the rest of the world (ROW). In defining these regions, we have attempted to employ the minimal amount of disaggregation necessary to provide meaningful insights into how the costs of a carbon constraint might vary among nations.

These regional categories have been based on two considerations. First, any solution to the climate problem is likely to require differentiated responses by industrialized and by developing countries. Second, over the long term, the CO$_2$ problem is primarily a coal problem, and nearly 90 percent of the world's coal resources are contained in the United States, Soviet Union, and China.

A distinction should be made between the first four regions and the ROW, a catchall category containing all countries not included in the other regions. It is required to maintain a consistent global balance of energy and carbon flows. However, the countries in ROW are expected to pursue their own individual interests rather than the welfare of the group as a whole. It would be misleading to treat the ROW as a homogeneous entity.

Within each region, the analysis is based on ETA-MACRO, a model of two-way linkage between the energy sector and the balance of the economy. This is a merger between ETA (a process model for energy technology assessment) and a macroeconomic production function that provides for substitution between capital, labor, and energy inputs. ETA-MACRO is a tool for integrating long-term supply and demand projections. It is designed to compare the options that are realistically available to each region as the world moves away from its heavy dependence on oil and gas resources and toward a more diversified energy economy. The basic concepts of the model have been well tested. (For an earlier version, see Manne 1981.)

ETA-MACRO simulates a market or a planned economy over time. There is a single representative producer-consumer. Supplies, demands, and prices are matched through a dynamic nonlinear programming model. The higher that prices rise, the greater the amounts of future supplies that are likely to become available, and the greater

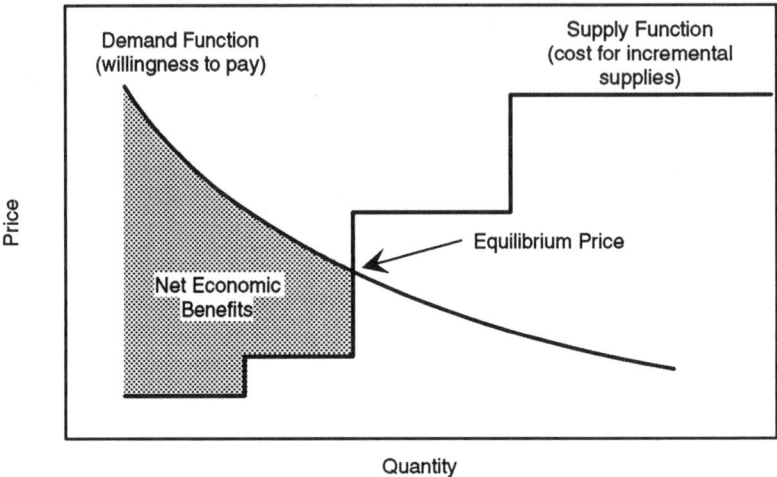

Figure 2.1 Market mechanisms and maximization

will be the inducements for consumers to conserve energy. Price responsiveness is lower in the short run than over the long term.

Figure 2.1 represents an example of partial equilibrium reasoning applied to a single energy form in a single time period. For illustrative purposes, consumers' willingness to pay is shown as a smoothly decreasing function of the amount of energy available to them, and producers' incremental costs are shown as a rising step function of the amount to be supplied. As in ETA-MACRO, these functions represent energy demands through a continuously differentiable production function and energy supplies through a stepwise linear physical process model. Supplies and demands are matched through an equilibrium price—one that just balances off the consumers' willingness to pay against the producers' incremental costs. It is as though the economy were attempting to maximize the size of the shaded area (net economic benefits).

In a partial equilibrium analysis, the supply and demand functions are independent of each other. By contrast, in ETA-MACRO, supplies and demands interact with each other through their impact on the GDP. Figure 2.2 provides an overview of the principal static linkages described within our model. Electric and nonelectric energy are supplied by the energy sector to the rest of the economy. Gross output depends on the inputs of energy, labor and capital. In turn, output is allocated among current consumption, investment in building up the

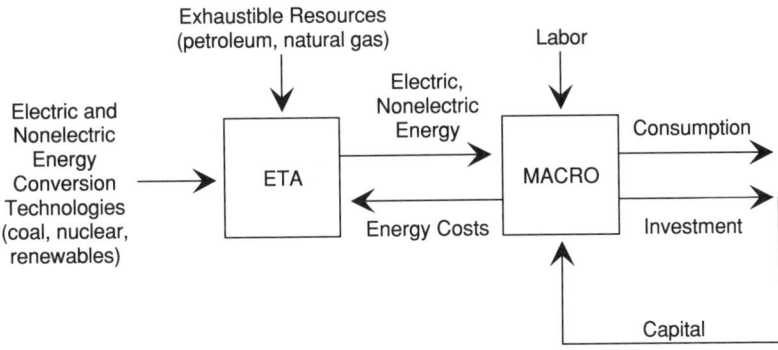

Figure 2.2   An overview of ETA-MACRO

stock of capital, and current payments for energy costs. (For a good exposition of the macroeconomic concepts employed here, see Allen 1968.)

We assume that carbon emission limits are imposed in such a way as to have a minimum impact on economic growth. For a market economy, this could be accomplished by auctioning off emission rights. Alternatively, one could limit emissions by imposing uniform carbon taxes upon the consumption of individual fuels.

It is supposed that the revenues from carbon taxes or from the sale of emission permits would be redistributed back to producers and consumers within the same region. We do not attempt to analyze international redistribution proposals, such as debt-for-nature swaps, and we do not attempt to analyze the distortions that might be created if, say, there were mandatory efficiency standards imposed on automobiles or other devices.

The model allows for the possibility of deferring carbon emissions from one time period to the next. This option plays a key role when the value of carbon rights is rising rapidly. It ensures that the value of these rights cannot rise at a more rapid rate than that determined by the marginal productivity of capital.

Global 2100 combines the five regional submodels into an integrated system. The degree of linkage varies depending on the issue under study. The most straightforward application is one in which oil is the only internationally traded commodity, and the international price of crude oil is determined outside the model. Each region faces a specified quota for its carbon emissions. We can then perform our computations in parallel for each of the five geopolitical regions.

These parallel computations provide useful insights, but they may lead to two types of inconsistencies: they do not ensure that the international supplies of oil will automatically match the demands, and they do not ensure that the value of carbon emission rights will be comparable from one region to another. In order to eliminate these inconsistencies, we need to integrate the region-by-region analyses.

In the case of international trade in crude oil, the model is almost consistent. It is assumed that the ROW region (which includes OPEC, the Organization of Petroleum Exporting Countries) sets an international price, the OECD nations are price takers, and the ROW meets their demands for net imports. Since oil is viewed as an exhaustible resource, the price will eventually move toward an upper bound determined by a backstop technology such as coal-based synthetic fuels. Iterative methods (such as alternative backstop dates) may be employed to eliminate the prospective gaps between oil supplies and demands, but it is sometimes awkward to apply these informal methods.

An international agreement on carbon emission limits might be based on many different criteria, and it would create a new form of property rights. There is no guarantee that a politically acceptable allocation would equalize the marginal value of emission rights among regions. In negotiating this type of agreement, it might be possible to separate the issue of equity from that of efficiency. To promote economic efficiency, the agreement could provide for international trade in carbon emission permits.

To analyze this type of trade through Global 2100, each region is viewed as a price taker and as a possible importer or exporter. Each is coupled to the others through the international price. Since this is an intertemporal problem, the time path of carbon prices must be determined so as to equilibrate supplies and demands during each period simultaneously. We can no longer formulate the overall problem as five independent nonlinear optimizations. Instead, this is viewed as an equilibrium problem in which all choices are integrated through an international market in emission rights.

Key Demand Assumptions

The rate of GDP growth is a key determinant of energy demands. This rate depends on both population and per capita productivity trends. In parallel with the slowing of population growth during the

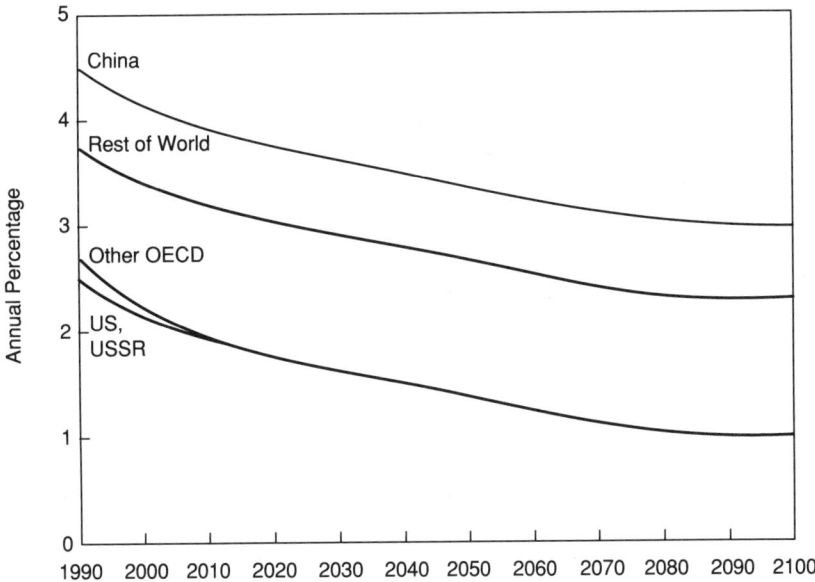

Figure 2.3 Potential GDP growth rates

twenty-first century, there will be a diminishing rate of growth of GDP and, hence, a slowdown in the demand for energy. Figure 2.3 shows our assumptions about the rate of potential GDP growth within each region. These rates represent the average of the higher- and lower-growth cases adopted by the IPCC's Working Group III (1990). Typically, they represent extrapolations of growth performance achieved during the 1970 through 1990 period. Appendix A documents these growth rates and all of the other macroeconomic parameters employed in our business-as-usual demand projections.

Because of energy-economy interactions, the potential GDP growth rates do not uniquely determine the realized rates. Energy costs represent just one of the claims on the economy's output. Tighter environmental standards and/or an increase in energy costs will reduce the net amount of output available for meeting current consumption and investment demands. The potential will then exceed the realized GDP.

Energy consumption need not grow at the same rate as the GDP. Over the long run, they may be decoupled. In Global 2100, these possibilities are summarized through two macroeconomic parameters: ESUB (the elasticity of price-induced substitution) and AEEI (autonomous energy efficiency improvements).

If there is sufficient time for the adaptation of capital stocks, most analysts would agree that there is a good deal of possible substitutability between the inputs of capital, labor, and energy. The degree of substitutability will affect the economic losses from energy scarcities and price increases. In our model, the ease or difficulty of these trade-offs is summarized by ESUB. The higher the value of ESUB is, the less expensive it is to decouple energy consumption from GDP growth during a period of rising energy prices.

When energy costs are a small fraction of total output, ESUB is approximately equal to the absolute value of the price elasticity of demand. In Global 2100, this parameter is measured at the point of secondary energy production: electricity at the busbar, crude oil and synthetic fuels at the refinery gate. The numerical values of ESUB have been taken to be .40 for the United States and other OECD. These countries have already demonstrated their ability to use the price mechanism as an aid in decoupling energy from GDP growth. Elsewhere, price-induced substitution is assumed to be more problematic, and it depends on structural economic changes. We have therefore set ESUB at .30 for these other three regions.

In addition to the reductions in energy demand induced by rising energy prices, there is also the impact of autonomous energy efficiency improvements. Nonprice efficiency improvements may be brought about by deliberate changes in public policy, such as speed limits for automobiles. Energy consumption may also decline as a result of shifts in the basic economic mix away from manufactured goods and toward more services. Thus, the AEEI summarizes all sources of reductions in the economy-wide energy intensity per unit of output.

Figure 2.4 shows the effect of alternative parameter values. With an annual AEEI rate of 1.5 percent, by the end of the next century the economy would require only 20 percent of the amount of energy that it would require with an AEEI of zero. This is entirely apart from the conservation that would be induced by rising energy prices in response to the depletion of conventional oil and gas resources. Clearly, the AEEI is important, and its numerical value is highly controversial.

Econometric investigations of the U.S. post-1947 historical record show no evidence for autonomous time trends of energy conservation (Brown and Philips 1989, Hogan 1988, and Jorgenson and Wilcoxen

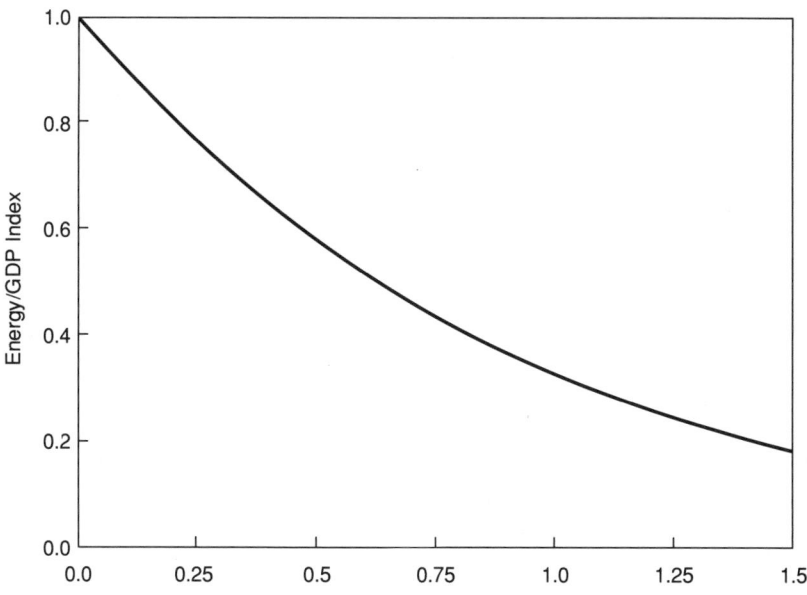

Figure 2.4 Effect of AEEI on primary energy consumption, 2100

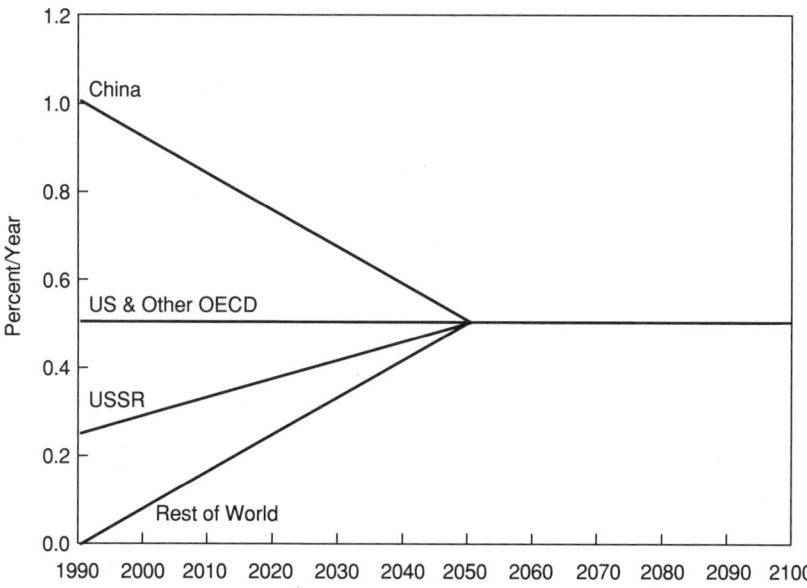

Figure 2.5 Autonomous energy efficiency improvements, AEEI

1989). Indeed, Hogan and Jorgenson (1991) suggest that the AEEI for the United States might have been negative during this period. Technologically oriented end-use analysts, however interpret the historical record quite differently. They attribute a significant role to nonprice efficiency improvements in the decoupling between energy and GDP growth rates (Goldemberg et al. 1987, and Williams 1990).

Figure 2.5 shows our AEEI assumptions. For the initial decades of the twenty-first century, the AEEI is 0.5 percent annually for the OECD countries, 0.25 for the Soviet Union, 0.0 for ROW, and 1.0 for China. The lower values for the Soviet Union and ROW reflect the fact that these regions will be undergoing further industrialization before moving toward a service-based economy. As a result, they are likely to experience fewer opportunities for reductions in their energy intensiveness than if they were further along toward a postindustrial phase. Recall that China has the highest economic growth targets among the five regions. Unless it makes more efficient use of its energy resources, achieving these GDP targets will become exceedingly difficult. The higher AEEI value for China (an annual rate of 1.0 percent) reflects high internal political pressures to decouple energy and economic growth. As other countries increasingly take on the postindustrial characteristics of the OECD nations, the AEEI differentials among regions are likely to decline. For the second half of the twenty-first century, we assume a uniform AEEI of 0.5 percent annually throughout the world.

In addition to the ESUB and AEEI, one other numerical parameter affects the decoupling between energy consumption and economic growth. The reference price of nonelectric energy, PNREF, is employed in benchmarking the demand functions for energy. Its numerical value is chosen so as to allow for differences in the extent to which domestic controls have insulated consumers from international energy price movements. As a consequence, the economies of some regions may not have completed the long-term adjustment to the oil price shocks of the 1970s. If PNREF is less than the actual 1990 base year oil price of $4 per GJ ($22 per barrel), there will be some near-future price-induced conservation in addition to that determined by the AEEI factor. In calculating these once-for-all efficiency gains, the value of PNREF is taken to be $3.50 per GJ in the United States, $4 per GJ in the other OECD, and $2 per GJ in the other three regions.*

---

* 1 GJ = 10^9 joules = $.95 \times 10^6$ British thermal units.

Key Supply Assumptions

The model distinguishes between electric and nonelectric energy. Appendixes A through E document all of the cost and performance characteristics attributed to these technologies. Each is characterized by its unit costs, fuel requirements, carbon emission coefficients, dates of introduction, and maximum rates of expansion or decline. The supply expansion constraints are not rigid upper bounds but soft constraints. Growth may be accelerated but at a rising marginal cost.

Table 2.1 identifies the alternative sources of electricity supply. The first five technologies represent existing sources: hydroelectric and other renewables, gas-, oil-, and coal-fired units, and nuclear power plants. The second group of technologies includes the new electricity generation options that are likely to become available. They differ in terms of their projected costs, carbon emission rates, and dates of introduction. These technologies are intended to be representative of existing and future options within the United States. Elsewhere, the cost and performance characteristics are adjusted for regional differences using judgment and first-order correction factors (Vejtasa and Schulman 1989).

It is expected that new gas-fired capacity for base load electricity will take the form of combustion turbine combined cycle plants—units that have a high thermal efficiency, low carbon emissions, and low capital costs. If natural gas prices remain at their 1990 levels, this technology would represent an attractive source of electricity; however, as natural gas resources gradually become exhausted, fuel prices are likely to rise. For example, in a baseline projection for the Gas Research Institute, Woods (1988) projects a tripling of U.S. wellhead prices by 2010. With an increase of this magnitude, gas-fired electricity would lose its competitive advantage over coal.

There are many emerging coal technologies with attractive cost and performance features. In table 2.1, these options are represented by a new pulverized coal plant with flue gas desulfurization. Its cost estimates are similar to those for advanced fluidized-bed combustion (AFBC) and integrated gasification combined cycle (IGCC) plants.

ADV-HC and ADV-LC, respectively, refer to advanced high- and low-cost carbon-free electricity generation. Any of a number of technologies could be included in these categories: solar, nuclear, biomass, and others. Given the enormous disagreement as to which of these

Table 2.1
Identification of electricity generation technologies

Technology name	Earliest possible introduction date[a]	Identification
Existing:		
HYDRO		Hydroelectric, geothermal, and other renewables
GAS-R		Remaining initial gas fired
OIL-R		Remaining initial oil fired
COAL-R		Remaining initial coal fired
NUC-R		Remaining initial nuclear
New:		
GAS-N	1995	Advanced combined cycle, gas fired
COAL-N	1990	New coal fired
ADV-HC	2010	High-cost carbon free
ADV-LC	2020	Low-cost carbon free

a. Estimated year when the technology could provide .1 trillion kWh (approximately 20 GW of installed capacity at 60 percent capacity factor).

will ultimately win out in terms of economic attractiveness and public acceptability, we have chosen to refer to them generically. To allow for technical progress in reducing costs over time, we assume that ADV-HC will become available ten years earlier than ADV-LC.

Table 2.2 identifies the nine alternative sources of nonelectric energy within the model. The list is headed by OIL-MX, imports less exports of crude oil. Petroleum is the international "swing" fuel and its price is crucial to any near- or medium-term projections of energy supplies and demands.

All other carbon-based fuels are ranked in ascending order of their cost per GJ of crude oil equivalent. The least expensive domestic source is CLDU: coal employed for direct uses in industries such as steel and cement. In most scenarios, its maximum potential growth rate is taken to be zero. Next in the merit order are domestic oil and gas. These domestic resources are available at a constant marginal cost but are subject to upper bounds on extraction rates based on a Hotelling-type model of reserves and resource depletion.

For determining the extraction rate of oil and gas, we draw a sharp distinction between current proved reserves and the remaining stock of undiscovered resources. We do not attempt to estimate continuously rising marginal cost curves. Instead, just two categories of oil and two categories of natural gas are distinguished: low and high cost. Proved

Table 2.2
Nonelectric energy supplies

Technology name	Description	Carbon emission coefficient (tons of carbon per GJ of crude oil equivalent)	Unit cost per GJ of crude oil equivalent (1990 Dollars)
OIL-MX	Oil imports minus exports	.0199	4.00 in 1990 rising to 8.40 from 2040 onward
CLDU	Coal—direct uses	.0241	2.00
OIL-LC	Oil—low cost	.0199	2.50[a]
GAS-LC	Natural gas—low cost	.0137	2.75[a,b]
OIL-HC	Oil—high cost	.0199	6.00
GAS-HC	Natural gas—high cost	.0137	6.25[b]
RNEW	Renewables	.0000	8.20
SYNF	Synthetic fuels	.0400	8.33
NE-BAK	Nonelectric backstop	.0000	16.67

Note: The source of most of these carbon emission and cost coefficients is Energy Modeling Forum 12 (1990).

a. Oil-LC costs are only $0.50 per GJ in ROW. Similarly, GAS-LC costs are only $1.00 per GJ in ROW.

b. To allow for gas distribution costs, $1.25 per GJ is added to the wellhead price.

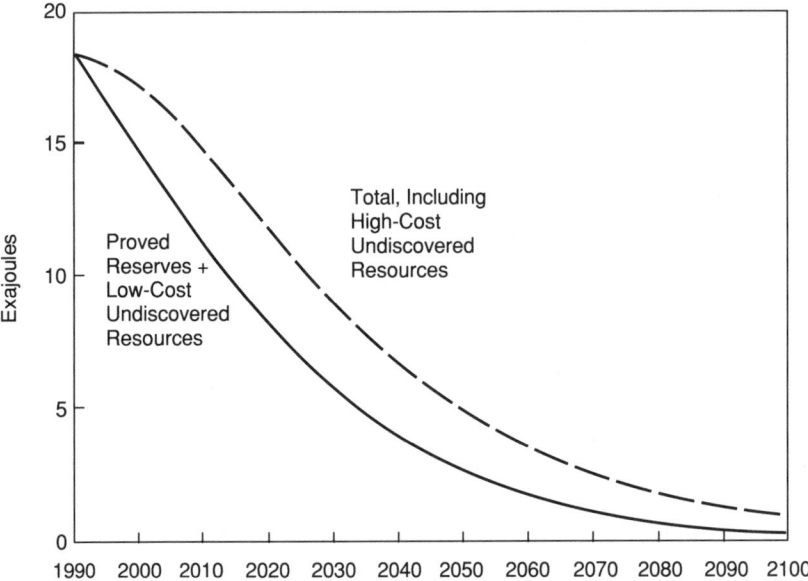

Figure 2.6 Natural gas (maximum U.S. production rates)

reserves are depleted by current production and augmented by new discoveries out of the remaining stock of undiscovered resources.

This is almost a constant ratio model of resource extraction. At any one time, production is a fixed fraction of remaining reserves. New discoveries may not exceed a fixed fraction of the remaining undiscovered resources. The only element of flexibility lies in the ability to defer reserve additions. Global 2100 is able to incorporate forward-looking resource-depletion policies. At the same time, the model is capable of representing an important real-world phenomenon. During a given year, a region may be importing oil—and also engaging in domestic production out of both low- and high-cost resources.

Figure 2.6 illustrates the maximum rate of U.S. domestic natural gas production, assuming that international oil prices rise at a sufficiently rapid rate so that there is no deferral of reserve additions. For purposes of this figure, the maximum resource depletion factor (RDF) is taken to be 5 percent per year. When the production-reserve ratio exceeds the RDF, it can be shown that the RDF governs the ultimate rate of production decline.

Our undiscovered resource estimates are taken from the ninety-fifth percentile point along the probability distributions available from the

U.S. Geological Survey work by Masters et al. (1987). These resources are split equally between those in the low- and high-cost categories. Masters et al. provide a modal (i.e., most likely) estimate of resources along with the fifth and ninety-fifth percentiles. For practical purposes, the ninety-fifth percentile point indicates an upper bound on undiscovered conventional resources, and the fifth percentile indicates a lower bound. That is, according to the USGS, there is a 95 percent probability that undiscovered conventional resources will not exceed the ninety-fifth percentile values.

By taking the USGS upper bound on conventional resources, we have attempted to allow for future technological progress, such as further reductions in the costs of deep drilling. Had our calculations been based on the modal or the fifth percentile, the prospects for conventional gas production would be considerably more pessimistic than the case examined here. In any case, figure 2.6 suggests that there is little likelihood that domestic resources will permit a significant expansion of U.S. consumption above the 1990 level. Additional gas consumption will require either pipeline or LNG imports from other regions.

With regard to carbon-free alternatives, the choices have been grouped into two broad categories: RNEW (low-cost renewables such as ethanol from biomass) and NE-BAK (high-cost backstops such as hydrogen produced through photovoltaics and electrolysis). The key distinction is that RNEW is in limited supply but NE-BAK is available in unlimited quantities at a constant but considerably higher marginal cost. In the absence of a carbon constraint, SYNF (synthetic fuels based on coal or shale oil) would place an upper bound on the future cost of nonelectric energy.

Data Issues

The global climate debate presents enormous challenges for economic modelers. Ideally, international negotiators would like to know how the costs of a carbon constraint might vary across regions. To address this question, we would have to analyze the existing energy-economic system of each nation and project how it is likely to evolve over time. It is difficult enough to describe what is going on around us today.

Even for the United States, with its enormous database, it is not easy to reach consensus on some of the most fundamental relationships underlying the existing energy system. A good example is the ongoing (and often heated debate) over the relative roles of price versus non-price mechanisms in the decoupling between energy and economic growth during the years since the 1973 oil price shock. The rate of nonprice-induced conservation may have ranged anywhere between zero and several percent a year. The difference is explained by alternative interpretations of the historical record. If analysts cannot reach agreement in the United States with its substantial base for econometric investigation, how can they be expected to agree in other nations where the evidence is much more limited?

There can even be difficulties in estimating something as straightforward as the base year GDP (conventionally measured in terms of the output of marketable goods and services). Seemingly, there are objective estimates available for these statistics. In practice, however, international GDP comparisons can be misleading. When they are based on official exchange rates and when they fail to allow for differences in purchasing power parity, they provide a poor indication of the relative income of a country. In assembling our base year statistics for China, we encountered GDP estimates that varied by a factor of five. (Compare Wolf et al. 1989 with Central Intelligence Agency 1989.) If we do not have a clear idea of base year levels, we are on shaky ground in making long-term projections of the percentage GDP losses associated with carbon emission constraints.

It is hazardous to project the cost and performance characteristics for emerging new technologies. By definition, these advanced technologies lie in the future. The present and recent past may tell us something about what to expect, but there is no sure way to predict which technologies will ultimately win out and how attractive they will turn out to be. We can make educated guesses about the future character of the energy sector, but these conjectures are subject to wide margins of error.

It is a daunting task to estimate CO_2 abatement costs, and any calculations will be highly speculative. Until key uncertainties can be reduced, we must be content to deal in broad ranges. In designing Global 2100, one of our objectives was to develop a model with which it would be easy to explore a wide series of assumptions. We begin this process with the United States, the country for which we have the best understanding of the current system and how it is likely to evolve. The data difficulties are compounded when we turn to other regions in the world.

Extensions and Suggestions for Additional Research

In Global 2100, the analysis of international oil markets is inherently heuristic. The model is almost consistent but not quite. There can be periods during which oil export demands would exceed the supplies available from the ROW. This means that OPEC could raise its price above the level employed here. It would be possible to avoid these inconsistencies and eliminate demand-supply gaps through an AGE (applied general equilibrium) modeling approach such as that of Perroni and Rutherford (1991).

The AGE format would have further advantages. Since it allows for multiple agents, each with its own objective function and budget constraints, it would enhance our ability to examine the impact of carbon limits on international trade. This could be particularly important in two sectors where trade is omitted from the current version of Global 2100: energy-intensive basic materials and natural gas. Carbon emission limits could have major impacts on the location of those industries producing energy-intensive basic materials. Pipeline movements of natural gas would be important in the case of exports from the Soviet Union to Europe and Japan. An AGE model would also permit us to analyze public finance issues—specifically the tax and expenditure aspects of alternative proposals for carbon limitations.

All of these topics deserve consideration, and none of them can be analyzed satisfactorily on the back of an envelope. AGE modeling would require additional data, and would push the limits of our computing capabilities. Pending these developments, it should be possible to extend Global 2100 to include a number of topics that are not covered within this book but could be included with modest amounts of additional effort. The existing modeling structure could be extended to allow for interactions between carbon dioxide and other greenhouse gases.

One could characterize several of the supply technologies through smooth upward-sloping supply curves rather than through the stepwise linear formulation employed here. Clearly this would improve our analysis of renewable sources of energy supply. It would be straightforward to introduce reforestation options as a means of CO2 disposal. All these ideas are worth exploring, but it is easier to draw up a list of topics than to perform the research work. Issues need to be examined in an orderly priority sequence. One must continually ask whether a specific model refinement is likely to affect one's over-all estimate of the costs of purchasing different types of greenhouse insurance.

References

R. G. D. Allen. 1968. *Macroeconomic Theory*. Macmillan, New York.

S. P. A. Brown and K. R. Phillips. 1989. "An Econometric Analysis of U.S. Oil Demand." Federal Reserve Bank of Dallas, January.

Central Intelligence Agency. 1989. *Handbook of Economic Statistics*. CPAS 89-10002. Central Intelligence Agency, September.

Electric Power Research Institute. 1989. *Technical Assessment Guide*. Electric Power Research Institute, Palo Alto, Calif.

Energy Modeling Forum. 1990. "First-Round Study Design for EMF 12." Stanford University, December.

J. Goldemberg, T. B. Johansson, A. K. N. Reddy and R. H. Williams. 1987. *Energy for a Sustainable World*. World Resources Institute, Washington, D.C.

W. W. Hogan. 1988. "Patterns of Energy Use Revisited." Harvard University, June.

W. W. Hogan and D. W. Jorgenson. 1991. "Productivity Trends and the Cost of Reducing CO_2 Emissions." *Energy Journal* 12, no. 1, January.

D. W. Jorgenson and P. J. Wilcoxen. 1989. "Environmental Regulation and U.S. Economic Growth." Harvard University, July.

A. S. Manne. 1981. "ETA-MACRO: A User's Guide." EA-1724. Electric Power Research Institute, Palo Alto, Calif.

C. D. Masters, E. D. Attanasi, W. D. Dietzman, R. F. Meyer, R. W. Mitchell and D. H. Root. 1987. "World Resources of Crude Oil, Natural Gas, Natural Bitumen, and Shale Oil." *12th World Petroleum Congress*, Proceedings, vol. 5.

National Academy of Sciences. 1991. *Policy Implications of Greenhouse Warming—Synthesis Panel*. National Academy Press, Washington, D.C.

C. Perroni and T. Rutherford. 1991. "International Trade in Carbon Emission Rights and Basic Materials: General Equilibrium Calculations for 2020." Wilfrid Laurier University and University of Western Ontario, April.

S. A. Vejtasa and B. L. Schulman. 1989. "Technology Data for Carbon Dioxide Emission Model: Global 2100." SFA Pacific, Mountain View, Calif.

R. H. Williams. 1990. "Low-Cost Strategies for Coping with CO_2 Emission Limits." *Energy Journal* 11 (4), October.

C. Wolf, G. Hildebrandt, M. Kennedy, D. P. Henry, K. Terasawa, K. C. Yeh, B. Zycher, A. Bamezai and T. Hayashi. 1989. "Long-Term Economic and Military

Trends, 1950–2010." N-2757-USDP. RAND Corporation, Santa Monica, Calif. April.

T. J. Woods. 1988. "The Long-Term Trends in U.S. Gas Supply and Prices: The 1988 GRI Baseline Projection of U.S. Energy Supply and Demand to 2010." Gas Research Institute, Washington, D.C., December.

Working Group III. 1990. "Formulation of Response Strategies." Intergovernmental Panel on Climate Change, Washington, D.C.

3

The Costs of Limiting U.S. CO_2 Emissions

Introduction

There is good reason to begin our analysis with the United States. It is the region with which we are most familiar and for which we possess the best data. As of 1990, it was the largest emitter of CO_2. Figure 3.1 shows 1990 carbon emissions for each of our five geopolitical regions. Approximately 6 billion tons of carbon were emitted to the atmosphere as a result of fossil fuel burning. Nearly 25 percent was due to the combustion of fossil fuels within the United States.

There have been numerous proposals for reducing CO_2 emissions. Typically, these call for a greater push toward energy efficiency and a shift away from carbon-intensive fuels. The proposals differ in their emphasis on price- and nonprice-induced conservation. Rising international oil prices, technical progress, structural changes, and the correction of market failures could all contribute to reducing energy consumption per unit of economic output.

To appreciate the opportunities for fuel switching in the United States, one must take a closer look at the composition of its energy sector. In 1990, the electric sector was responsible for 37 percent of total carbon emissions. Figure 3.2 shows electric and nonelectric energy use by fuel type in that year. Coal-fired power plants supplied over 50 percent of electricity demands in 1990. In the short term, there may be opportunities for switching away from coal toward natural gas with its lower carbon emissions per unit of energy. For the longer term, there could be greater emphasis on carbon-free alternatives such as solar, fission, and fusion.

In the area of nonelectric energy use, oil is the principal fuel. If significant limitations are placed on CO_2 emissions, alternative supply sources must be found. Renewables (such as ethanol from biomass)

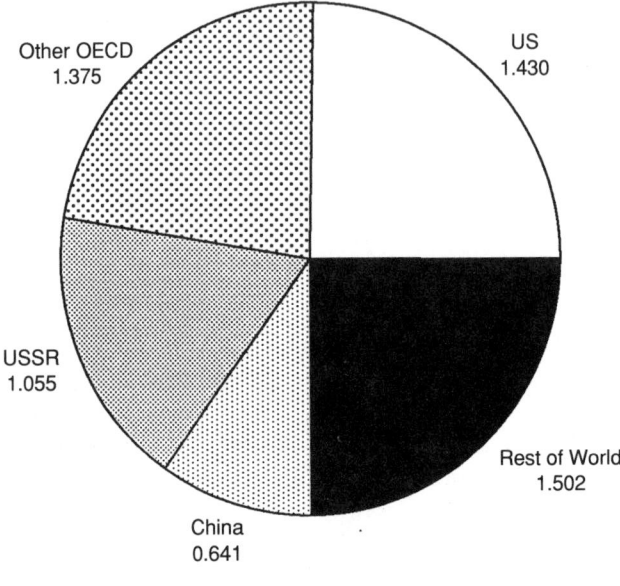

Billion Tons of Carbon

Figure 3.1 Carbon emissions from the combustion of fossil fuels, 1990

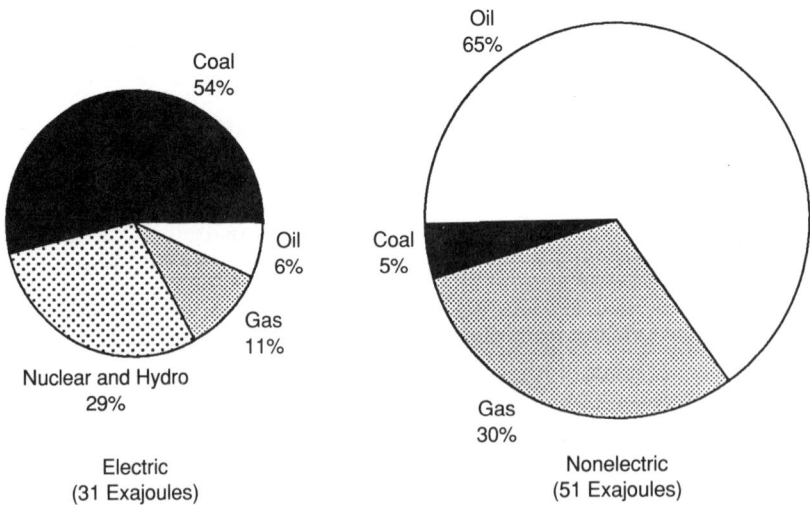

Figure 3.2 Primary energy consumption, 1990

may become economical but will remain in limited supply over the foreseeable future. High-cost backstops (such as hydrogen produced through photovoltaics and electrolysis) could be available in much larger quantities but at a considerably higher cost. More electrification is also an alternative. For example, if electric vehicles were charged with power from carbon-free electric generating facilities, CO_2 emissions could be reduced significantly.

The ease or difficulty of achieving an emissions reduction target will depend not only on the current composition of the energy sector but also on how the energy sector is likely to evolve in the absence of carbon limits. Would a business-as-usual energy future mean continued heavy dependence on carbon-intensive fuels? Alternatively, would the price mechanism lead toward nuclear power, carbon-free renewables, and/or energy efficiency? In this chapter, we explore a broad range of views about the future character of the energy sector. In each instance, we examine how demands for both electric and nonelectric energy are likely to evolve. We look at the role of coal and other carbon-intensive fuels in meeting those demands and see how this might affect the costs of carbon limits.

A Business-as-Usual Energy Future

We begin by projecting the evolution of the energy sector under business-as-usual conditions. (For details on our numerical assumptions, see chapter 2 and appendixes A through E.) Figure 3.3 presents successive snapshots of the electric and nonelectric sectors at four points in time. 1990 is based on the historical record. We measure electricity in terms of TkWh (trillion kilowatt-hours). Nonelectric and primary energy are expressed in exajoules (10^{18} joules). To within 5 percent, an exajoule is equivalent to one quad—that is, 10^{15} British thermal units. Economic quantities are valued in U.S. dollars of constant 1990 purchasing power. The figure highlights the importance of coal to the U.S. electricity industry. On the supply side are two principal alternatives to coal: natural gas and ADV-LC. If natural gas prices remain at 1990 levels, gas-fired electricity would represent an attractive option. However, geological resource constraints and competing demands from the nonelectric sector are likely to lead to significant gas price increases, a view broadly consistent with the estimates that appear in Woods (1988). In his baseline projections for the Gas Research Institute, Woods projects a tripling of wellhead prices by 2010.

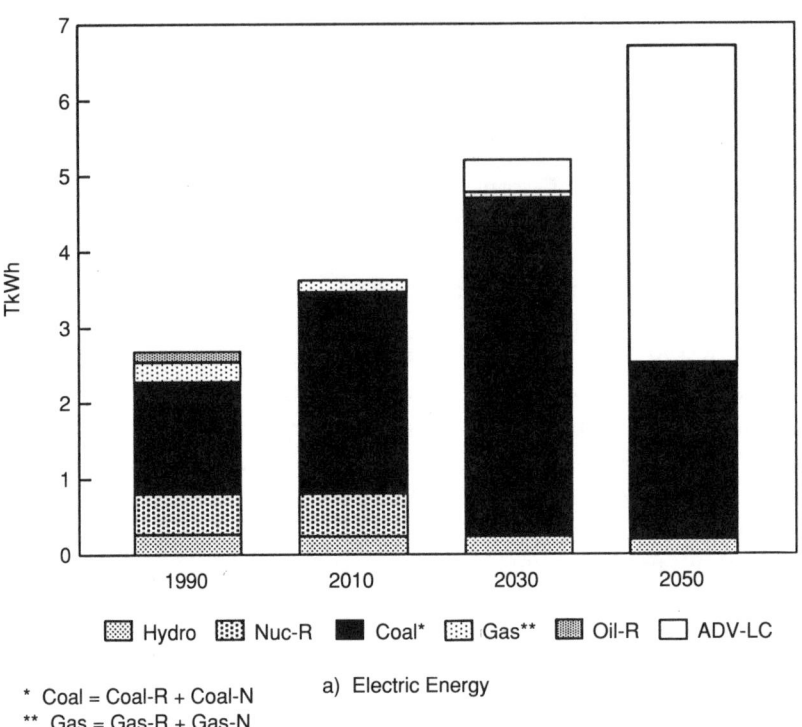

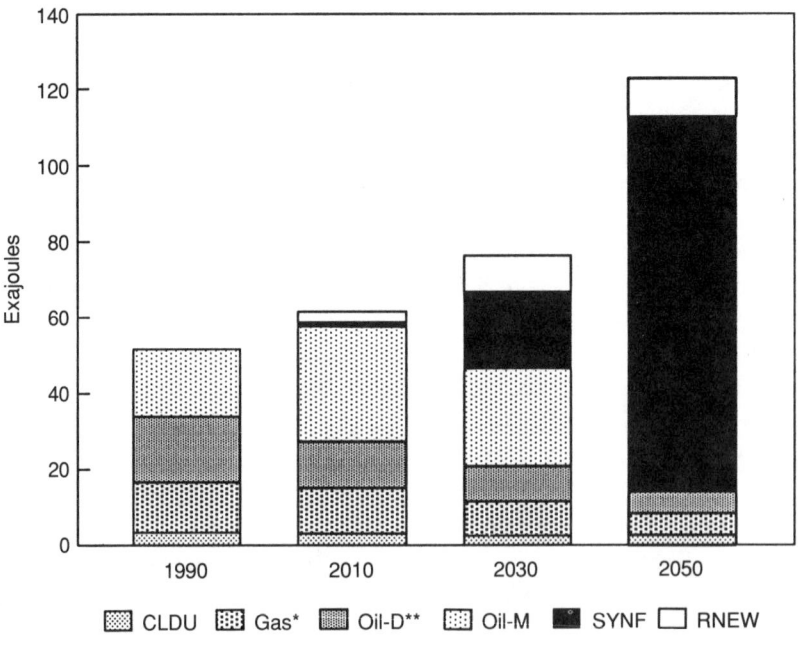

Figure 3.3 Energy sector under business as usual

With such an increase, gas-fired electricity would lose its competitive advantage over coal.

The other low-cost alternative to coal is the carbon-free technology, ADV-LC. Recall that any of a number of technologies could be included in this category: solar, nuclear, biomass, and others. We assume that at least one of these options will be attractive for economic reasons alone. We also assume that there are constraints on the rate at which it can enter the marketplace. If it were introduced in 2020, it would take on an increasing share of the electric load thereafter. This automatically leads to a substantial reduction in the rate of growth of carbon emissions.

On the nonelectric side, the story is more complicated. Through 2030, oil imports are the largest source of supplies. Since we assume that global oil and gas resources are limited, the exhaustion of these conventional fuels will lead to a run-up in nonelectric energy prices. As crude oil prices rise, new energy sources become attractive. These sources are grouped into two broad categories: SYNF (coal- and shale-based synthetic fuels) and RNEW (low-cost carbon-free renewables such as ethanol from biomass). With RNEW supplies limited to 10 exajoules per year, SYNF would eventually displace imported oil and become the marginal source of nonelectric energy.

A long-term increase in energy prices will affect not only supplies but also demands. Consumers will be motivated to cut back on their consumption of energy. Since ESUB (the elasticity of price-induced substitution) is .4, this means that a 1 percent price increase will lead to a decline of 0.4 percent in the demand for energy.

Figure 3.4 shows the emissions time path corresponding to our business-as-usual scenario. In the early decades (1990–2030), most of the increase is due to the addition of new coal-fired electricity plants. After 2030, the emissions growth is concentrated in the nonelectric sector. Coal-based synthetic fuels (the long-term marginal source of supply) are more carbon-intensive than those that they have replaced (crude oil and natural gas). Conversely, in the electric sector, coal is eventually replaced with carbon-free technologies. By the end of the twenty-first century, no carbon is released by the generation of electricity.

Despite the absolute growth in emissions, there would still be a slowdown in the percentage annual growth rate under business as

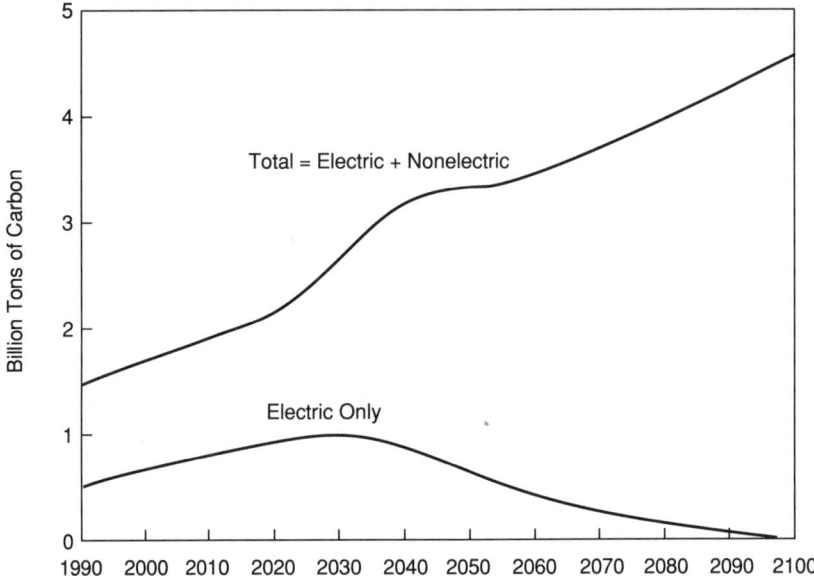

Figure 3.4 Carbon emissions under business as usual

usual. Between 1950 and 1990, U.S. emissions grew at an average annual rate of 1.8 percent. The growth rate during the twenty-first century is projected to be only 1.1 percent per year. Figure 3.5 is useful in
understanding the reasons for this slowdown. Part of the explanation
is a decline in the GDP growth rate to 1.5 percent annually. Although
this compounds to a fivefold increase during the twenty-first century,
the annual rate is low by historical standards. According to our calculations, carbon emissions and TPE (total primary energy) will grow
even more slowly than the GDP, for two reasons: fuel switching and
reductions in energy use per unit of output. The contribution of each
is indicated by figure 3.5. Note that fuel switching does not play a
role in reducing carbon per unit of economic output until after 2040.
Only then are carbon-free electric generating technologies available at
a sufficiently large scale to make an observable difference.

Impacts of a Carbon Constraint on the Energy Sector

Carbon constraints could take a variety of forms. Some legislative proposals have aimed only at slowing the rate of growth of emissions;

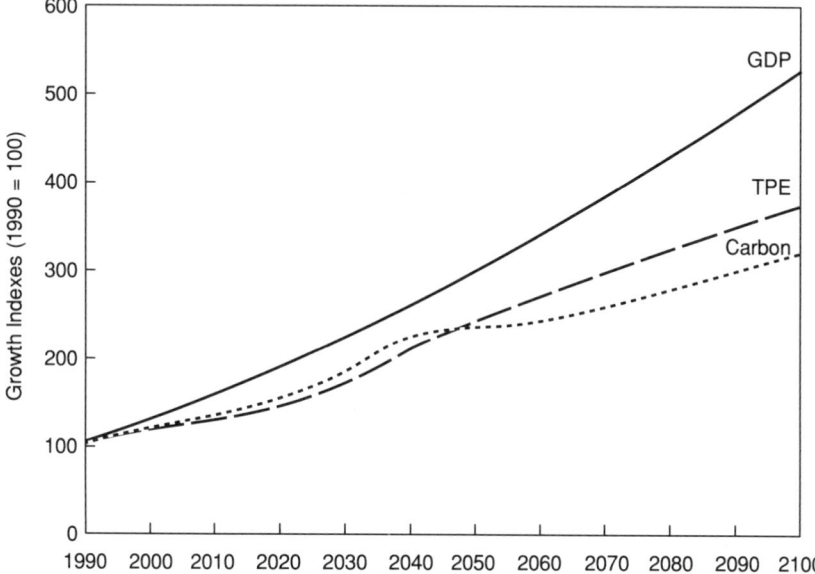

Figure 3.5   Growth indexes under business as usual

others have set targets of reducing emissions to half their current levels. Because of the wide range of options under consideration, Global 2100 has been designed with considerable flexibility regarding the imposition of carbon constraints.

We begin by investigating the costs of restricting emissions to 1.43 billion tons (their 1990 rate) through 2000, reducing them to 80 percent of this level by 2010, and stabilizing them thereafter. Although these targets are not as stringent as those contained in some proposed legislation, they nevertheless represent a substantial reduction in future emissions when compared with a business-as-usual view.

Figure 3.6 shows GDP, TPE, and carbon projections under this 20 percent emissions cutback. Over time, there is a change in the respective roles of conservation and of fuel switching. During the early decades (1990–2020), noncarbon-based supply options are severely limited. Conservation plays the dominant role. Through higher energy prices, GDP and energy growth are virtually decoupled. During the later decades, there are fewer constraints on the rate of market penetration of new sources of energy supply. Once the economy overcomes the introduction limits on carbon-free backstops, there is scope

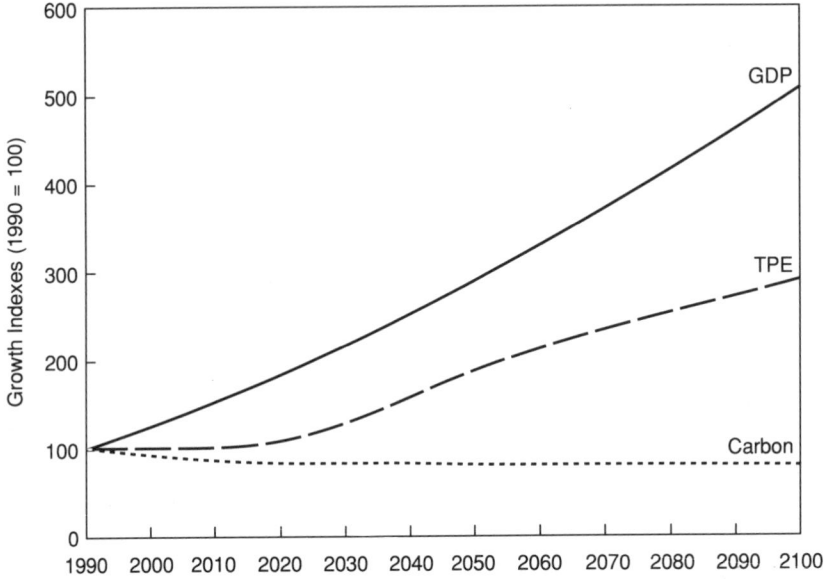

Figure 3.6 Growth indexes under a 20 percent reduction in carbon emissions

for a modest growth in energy consumption. Fuel switching is then the principal means of adapting to the carbon constraint.

Figure 3.7 indicates the importance of demand-side adjustments between 1990 and 2020, the period during which the economy is likely to have the most difficulty in adjusting to a 20 percent reduction in carbon emissions. Under business as usual, the GDP and energy consumption grow at annual rates of 2.1 percent and 1.2 percent, respectively. The AEEI accounts for more than half the difference between these two growth rates, with the remainder due primarily to price-induced conservation. It is our estimate of how consumers will respond to the higher prices of oil and natural gas as these resources undergo gradual depletion in the United States and abroad.

With a 20 percent emissions cut, the price-induced effects become more pronounced. Energy and GDP are virtually decoupled. The energy-GDP ratio declines by more than 1.5 percent per year. Of this total, the AEEI accounts for 0.5 percent, and price-induced conservation for over 1 percent annually.

Figure 3.8 helps to explain the role of fuel switching. Our model allows directly for interfuel substitution. Electric and nonelectric energy are substitutes in many markets. For example, residential oil burners

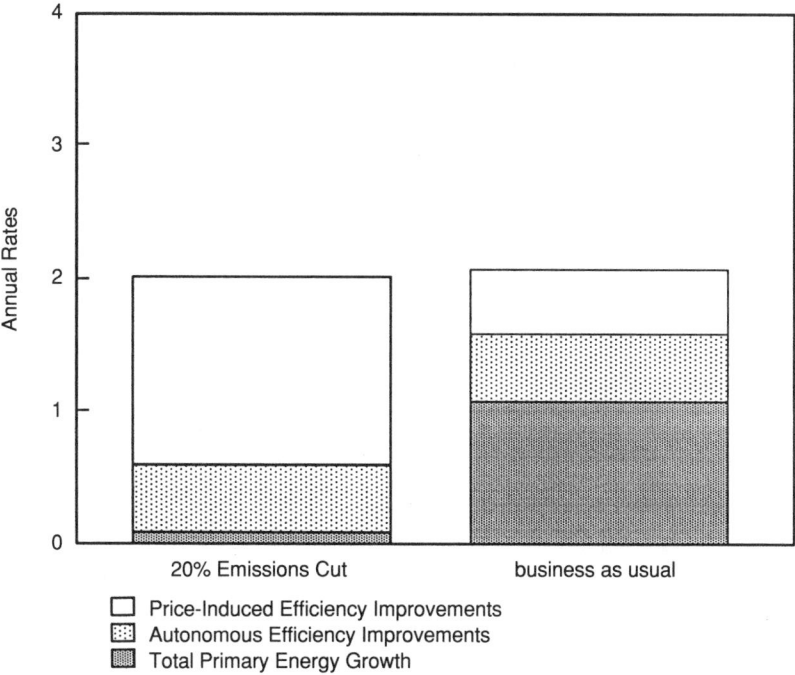

Figure 3.7 Decoupling between energy consumption and GDP growth, 1990–2020. Note: Components add up to rate of GDP growth.

may be replaced by gas burners, but they may also be replaced by electric heat pumps. If the price of one fuel rises relative to the other, there will be interfuel substitution. If emissions are to be reduced by 20 percent, figure 3.8 indicates that there would be a sharp run-up in the price of both electric and nonelectric energy between 1990 and 2020 but that their relative prices would remain roughly constant. After 2020, with the relaxation of the introduction constraints on ADV-LC, electricity prices begin to fall, but nonelectric energy prices continue to rise toward the backstop level. From 2020 on, electrification provides an economical way to reduce the energy system's reliance on carbon.

Figure 3.9 compares the share of total primary energy consumption devoted to the generation of electricity, with and without a carbon limit. Under business as usual, electricity's share remains roughly constant over time. With a carbon limit, electricity's share increases to nearly 60 percent by the end of the twenty-first century.

Figures 3.10 and 3.11 show the composition of the energy sector in 2010 and 2030 under the base case supply and demand assumptions.

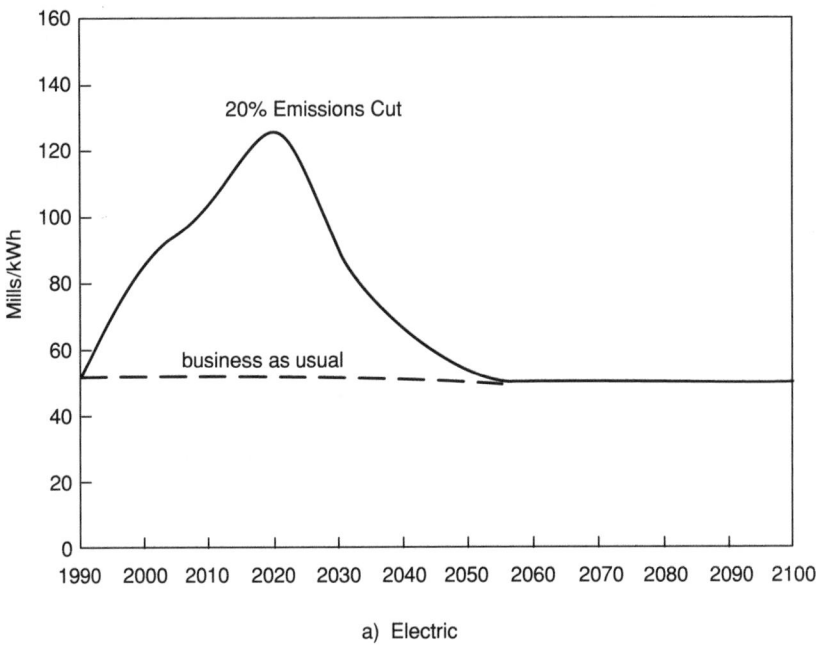

a) Electric

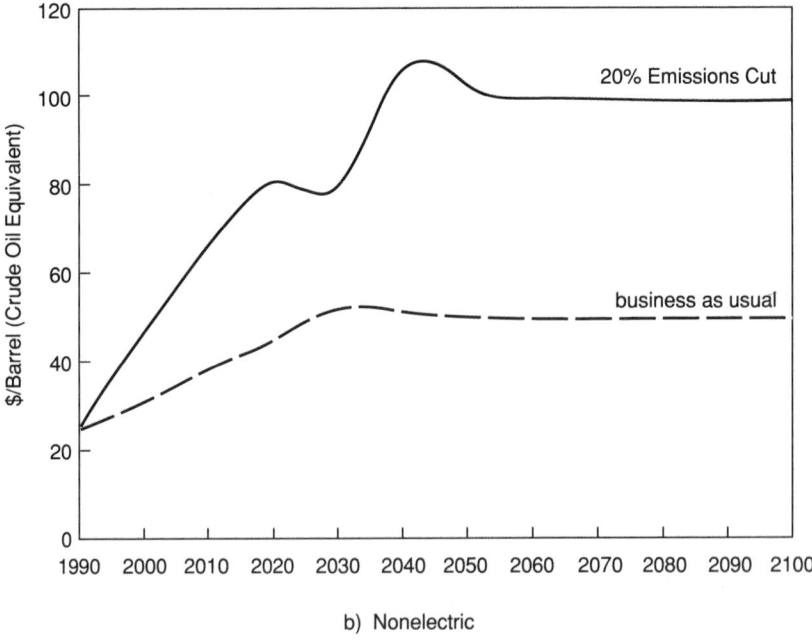

b) Nonelectric

Figure 3.8 Energy prices

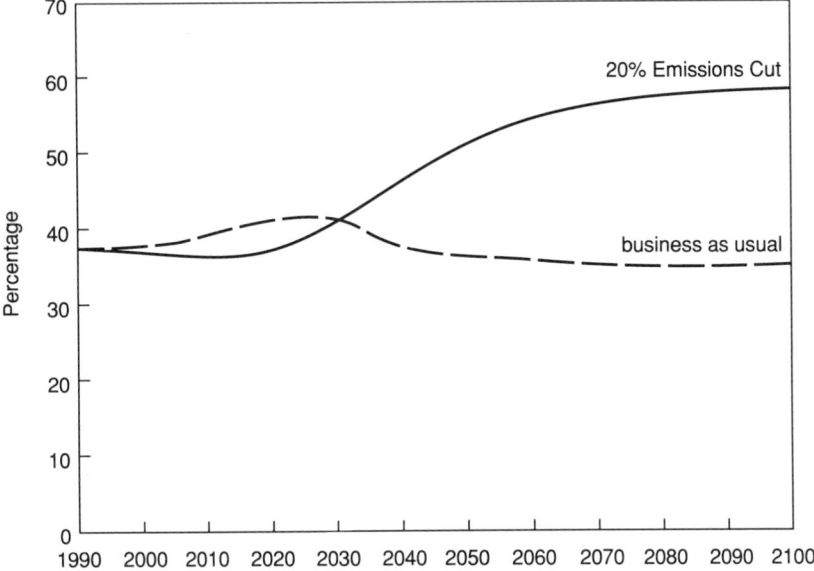

Figure 3.9 Electricity's share of total primary energy consumption

In each instance, the mix of supply alternatives is indicated, with and without the carbon constraint.

In the short term, there are severe constraints on the supply-side options for meeting electricity demands. Figure 3.10 indicates a significant rise of natural gas use within the electricity sector. Gas-fired plants produced only 10 percent of total electricity in 1990. According to Global 2100, the share of gas will increase to 27 percent by 2010 if we are in a carbon-constrained environment.

Eventually, increased demands will place tremendous pressures on natural gas markets. By 2010, the price of natural gas approaches the point at which high-cost advanced electricity technologies (ADV-HC) become attractive. Despite its cost, there is a significant role for this supply category during the period when the low-cost carbon-free alternatives (ADV-LC) are limited in their rate of market penetration. The imposition of a carbon constraint accelerates the introduction of both advanced carbon-free technologies. By 2030, they supply roughly 70 percent of electricity generation.

Figure 3.11 shows the impact of a 20 percent emissions cut on the nonelectric sector. The constraint implies a high penalty on carbon-intensive fuels. Crude oil and coal-based synthetic fuels become less

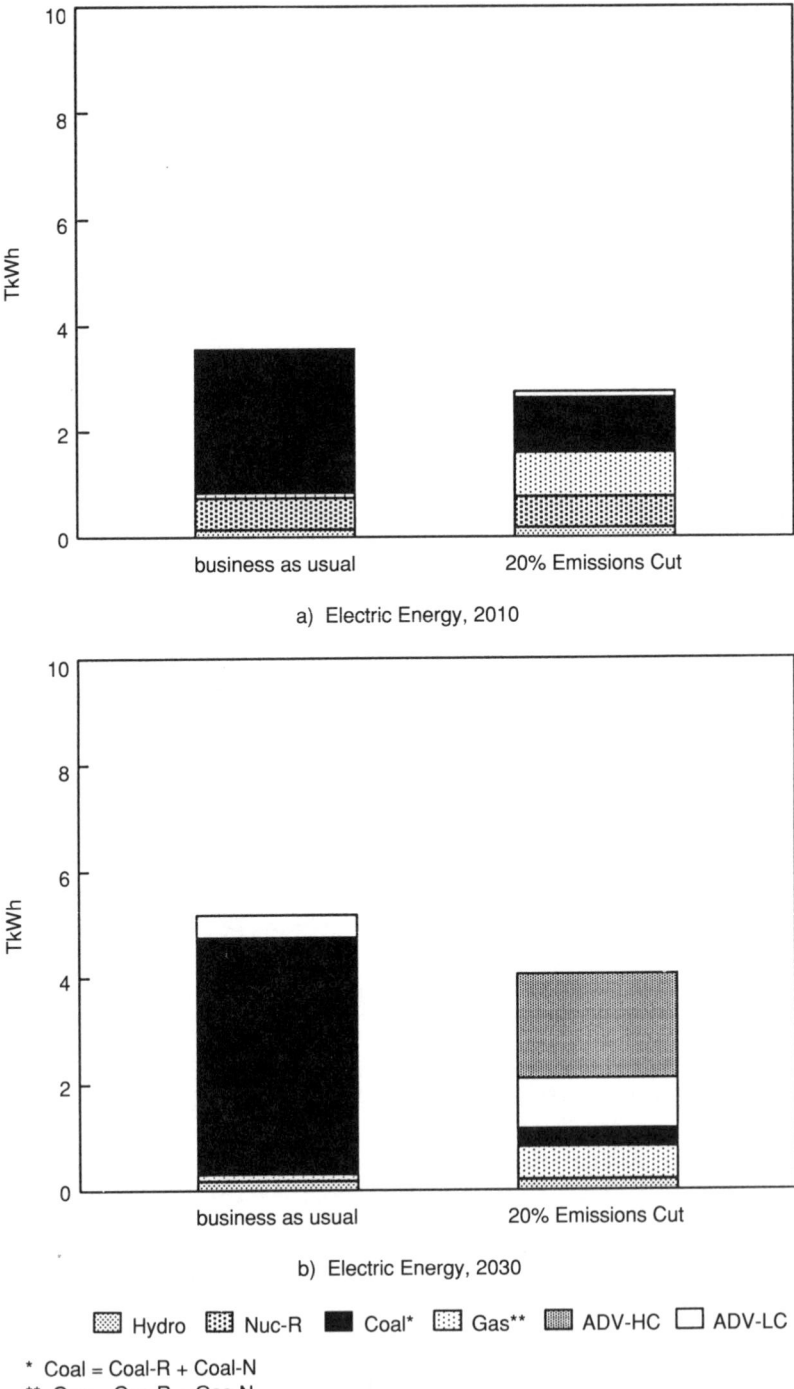

Figure 3.10 Electric energy projections

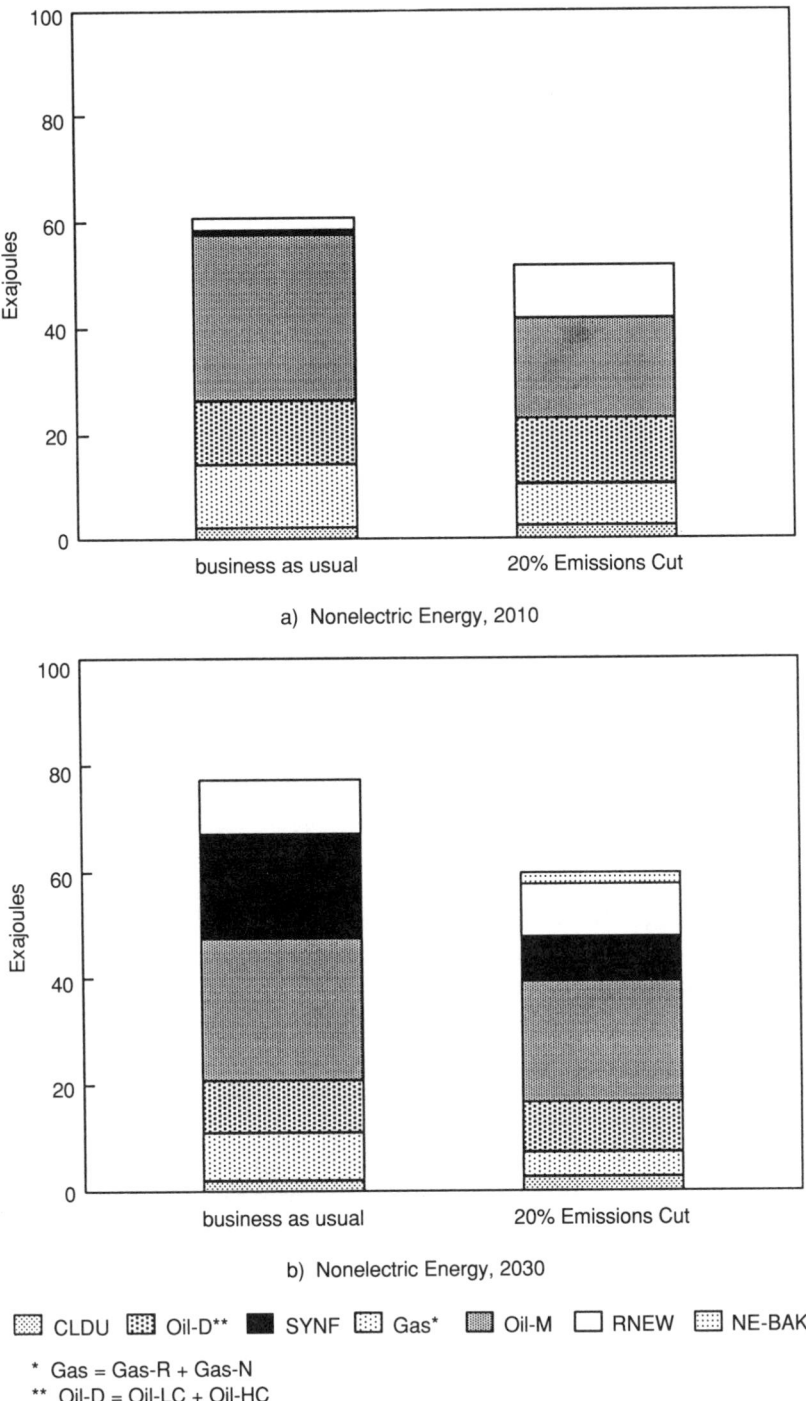

Figure 3.11 Nonelectric energy projections

desirable as sources of nonelectric energy. With the increased demand for natural gas for electricity, fewer gas resources are available for non-electric purposes. By 2000, nonelectric energy prices rise to the point where carbon-free renewables become attractive. If renewables were available in unlimited quantities, nonelectric prices would be capped at this point. In the base case, it is assumed that there are only limited supplies of renewables available at a low cost. Nonelectric energy prices would then continue to rise to the point where it becomes profitable to introduce the high-cost, carbon-free backstop, NE-BAK.

Economy-Wide Costs and Carbon Taxes

Using Global 2100, we add together the costs throughout the economic system and calculate the annual losses in GDP due to a carbon constraint. Figure 3.12 shows how these might vary over time. By 2000, the costs approach 1 percent of conventionally measured GDP. At that point, the rise in energy prices begins to have a significant effect on the share of gross output available for consumption and investment. By 2020, the GDP losses exceed 2 percent. Thereafter, this percentage grows to 2.5 percent and then remains constant. Adding over all the years from 1990 through 2100, the present value of the consumption losses would be $1.4 trillions, discounting to 1990 at 5 percent per year.

A variety of policy instruments are available for reducing carbon emissions to the specified levels. One option would be to impose a uniform tax per ton of carbon emissions. Figure 3.13 shows the size of the tax that would be required to induce consumers to reduce their dependence on carbon-intensive fuels. The tax begins at $135 per ton of carbon in 2000 and then rises sharply as emissions are reduced. In the absence of low-carbon supply alternatives, consumers are willing to pay a high price to burn carbon-based fuels. The tax must be sufficiently high to discourage these demands.

In most of our scenarios, it turns out that there are positive production levels for carbon-based synthetic fuels (SYNF) and for the carbon-free nonelectric backstop (NE-BAK). The long-run equilibrium tax rate must therefore be determined so as to make these supply technologies equally attractive to energy consumers. Specifically, the equilibrium rate is determined by the ratio of their cost differential to their carbon differential:

$$\frac{\text{Cost differential, \$/GJ}}{\text{Carbon differential, tons/GJ}} = \frac{\$16.67 - 8.33}{.04 \text{ tons}} = \$208/\text{ton}$$

A carbon tax of \$208 per ton would imply a fourfold increase in the price of coal—or an increase of \$.62 per gallon of refined petroleum products. According to figure 3.13, there are significant variations in the rate of the carbon tax at different times. First it overshoots and then undershoots its long-run equilibrium level. We cannot expect that a constant tax rate will be consistent with a level of emissions that remains constant over time.

Sensitivity of Costs to Supply- and Demand-Side Assumptions

Our base case supply and demand assumptions represent a view of the world in which carbon emissions continue to grow but at a slower rate than in the past. Table 3.1 compares this case with two alternative views about the potential for supply- and demand-side improvements in the energy sector. These differ with respect to the quantity

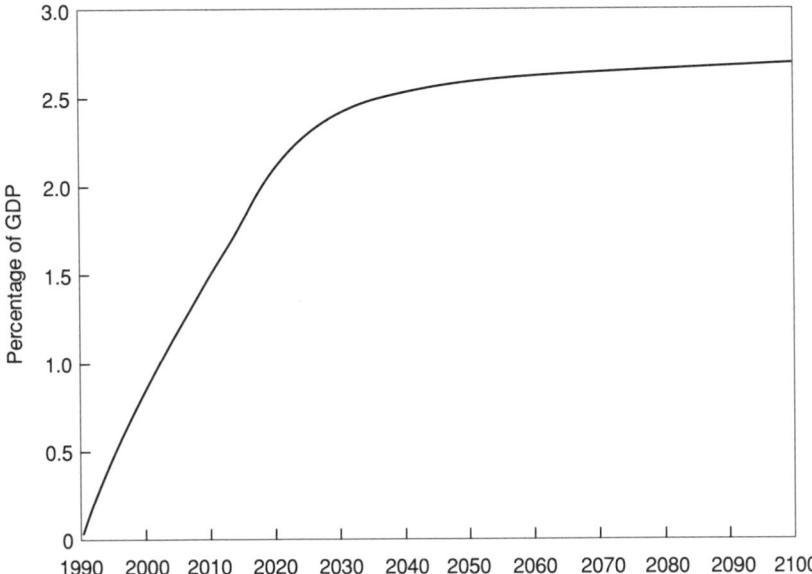

Figure 3.12 Annual losses due to carbon limit

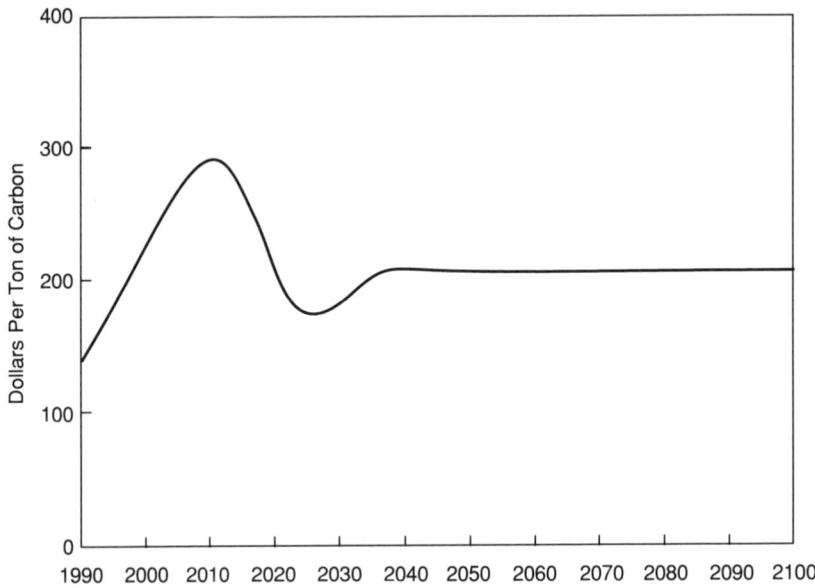

Figure 3.13 Carbon tax rates (value of emission rights)

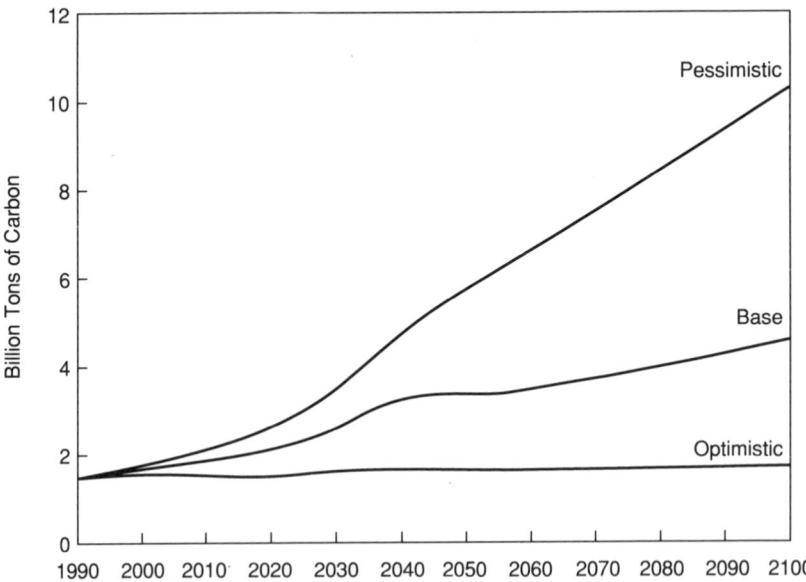

Figure 3.14 Alternative emissions scenarios: no carbon limits

Table 3.1
Alternative supply and demand scenarios

Scenario	RNEW (exajoules/year)	ADV-LC (mills/kwh)	AEEI (%/year)
Pessimistic	0	75	0.0
Base	10	50	0.5
Optimistic	20	50	1.5

available of RNEW (low-cost renewables), the cost of ADV-LC (the low-cost advanced electric supply technology), and the AEEI (the rate of autonomous energy efficiency improvement). The optimistic scenario describes an energy future that is considerably more attractive than the base case. Low-cost, carbon-free supply alternatives are in greater supply, and there is a lower overall demand for energy. The pessimistic scenario describes a very different future—one with fewer options for energy supply and conservation.

Figure 3.14 shows the implications for emissions when no carbon limits are imposed. Under the optimistic scenario, conservation and fuel switching combine to stabilize emissions at just a small amount above 1990 levels. Under the pessimistic scenario, carbon emissions grow considerably. In part, this is due to a higher energy intensity per unit of economic output and a lower availability of renewables. Even more important is the increase in costs of carbon-free electricity. The ADV-LC technology is no longer the marginal source of supply in the electric sector. In the pessimistic scenario, almost all new electricity demand is met by coal-fired power plants.

It turns out that the economic costs of carbon limits are quite sensitive to the assumptions about the potential for supply enhancements and demand conservation. If one subscribes to the pessimistic view of the future, the costs can be considerably higher than those presented for our base case. Figure 3.15 compares the pessimistic scenario with the two others. High demands and greater dependence on carbon-intensive fuels combine to increase the consumption losses to over $3.5 trillions (discounted to 1990 at 5 percent per year).

With the optimistic set of projections, an AEEI of 1.5 percent is sufficient to ensure that nonelectric energy demands are reduced to the level at which they can be covered eventually by 20 exajoules per year of RNEW, the low-cost renewable supply technology. Since we

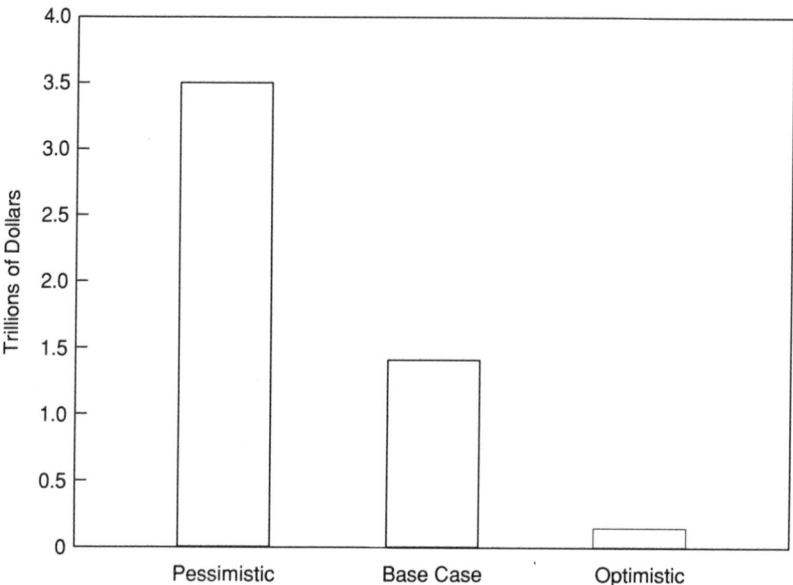

Figure 3.15 Discounted consumption losses, 20 percent emissions reduction

have assumed that RNEW is competitive with the coal-based alternatives and since no costs are imputed to autonomous energy efficiency improvements, there are negligible losses from a carbon constraint.

Figure 3.16 compares the time path for the carbon tax under all three scenarios. For the pessimistic case, carbon-free supply alternatives are significantly more expensive. Higher prices are required to induce consumers to reduce their dependence on carbon intensive fuels. The tax exceeds $600 per ton in 2010, the year in which the full 20 percent reduction takes effect. Eventually, the nonelectric backstop (NE-BAK) becomes available in sufficient quantities to put a $208 cap on the tax. This does not happen, however, until well into the twenty-first century.

By increasing the availability of RNEW, we delay the date at which the backstop becomes the marginal source of supply. Under the assumption of constant marginal costs, it is always economic to push the low-cost renewables to their limit before turning to NE-BAK, the high-cost nonelectric backstop. With an AEEI of 1.5 percent, the high-cost backstop is never needed, and it becomes irrelevant to the determination of the equilibrium carbon tax. The value of emission rights remains well below $208 per ton.

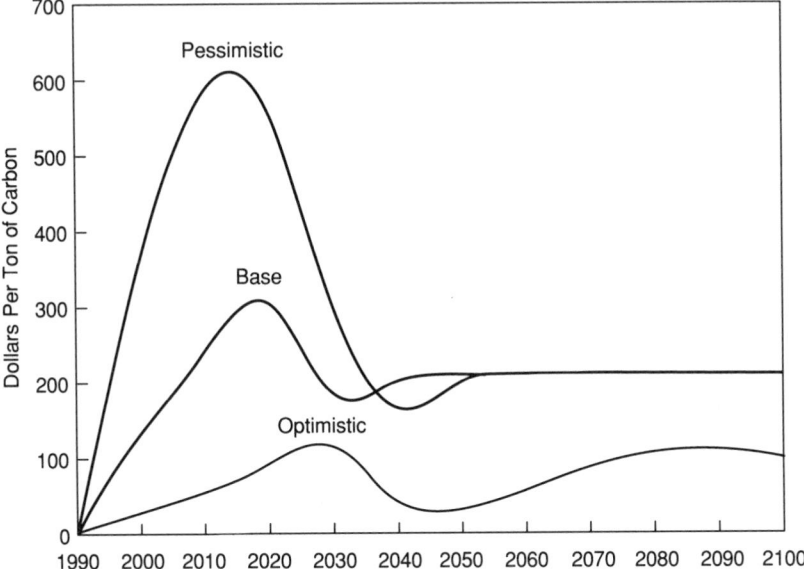

Figure 3.16 Carbon taxes: a sensitivity analysis

Sensitivity of Costs to an Alternative Carbon Emissions Constraint

Our final sensitivity analysis refers to an alternative carbon emissions constraint. Up to this point, we have considered only the case of reducing carbon emissions by 20 percent below 1990 levels. We began with this target because of its prominence in a number of proposals, both domestic and international. Proponents of more drastic controls point out that a 20 percent reduction would slow the rate of growth of cumulative emissions, but that because of the long lifetime of CO$_2$ in the atmosphere, this would not be sufficient to stabilize concentrations. If the goal is to stabilize atmospheric concentrations, more significant reductions will be required (Houghton et al. 1990).

To analyze scenarios of this type, we examine an even deeper reduction in emissions than a 20 percent cut: reducing emissions 50 percent below 1990 levels by 2010 and holding them at that level thereafter. Figure 3.17 compares the costs of alternative targets under the three supply-demand scenarios. For our base case, the tighter carbon constraint roughly doubles the abatement costs—from $1.4 trillion to $2.9 trillion (discounted to 1990 at 5 percent per year). At first glance this result may seem counterintuitive. Why isn't it more expensive to move

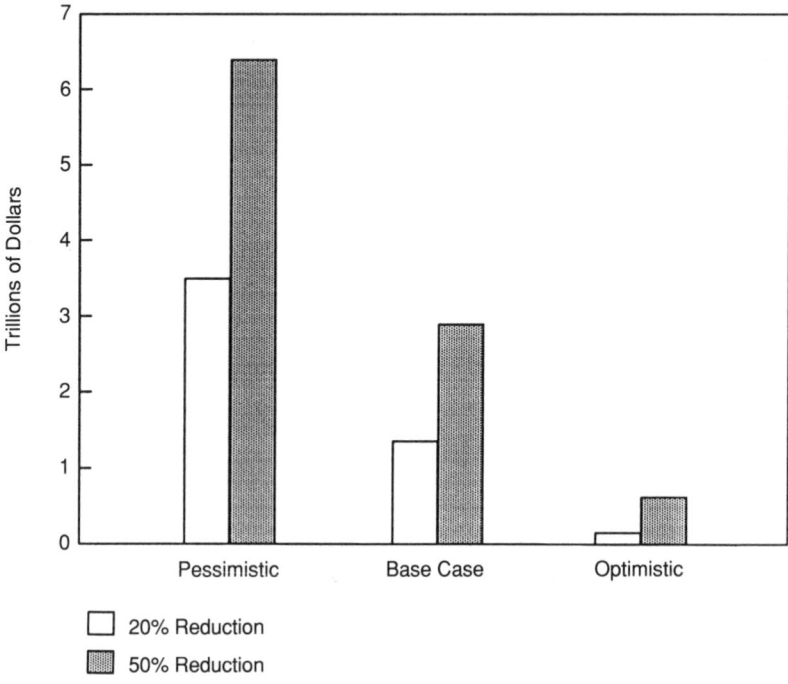

Figure 3.17 Discounted consumption losses under alternative carbon limits

from a 20 percent to a 50 percent target? After all, we have increased the size of the reduction by a factor of 2.5, and it is increasingly difficult to remove each additional ton of carbon from the energy system.

Figure 3.18 is helpful in understanding what is going on. It compares three emission profiles under base case supply and demand assumptions: business as usual, a 20 percent cutback below 1990 levels, and a 50 percent cutback below 1990 levels. In the absence of abatement measures, emissions continue to grow. This means that if we cap emissions at 20 percent below 1990 levels, the size of the required cutback will also grow over time. By 2100, emissions are 75 percent below what they would have been in the absence of the cap.

If emissions are reduced still further (to 50 percent below 1990 levels), in 2100 they will be 84 percent below what they would have been in the absence of the constraint. Note, however, that most of the reduction is achieved by the 20 percent cutback. Going from 20 percent to 50 percent has a relatively small effect on the absolute size of the reduction but a significant impact on the costs. It is increasingly expensive to remove each additional ton of carbon. This is why it is just as

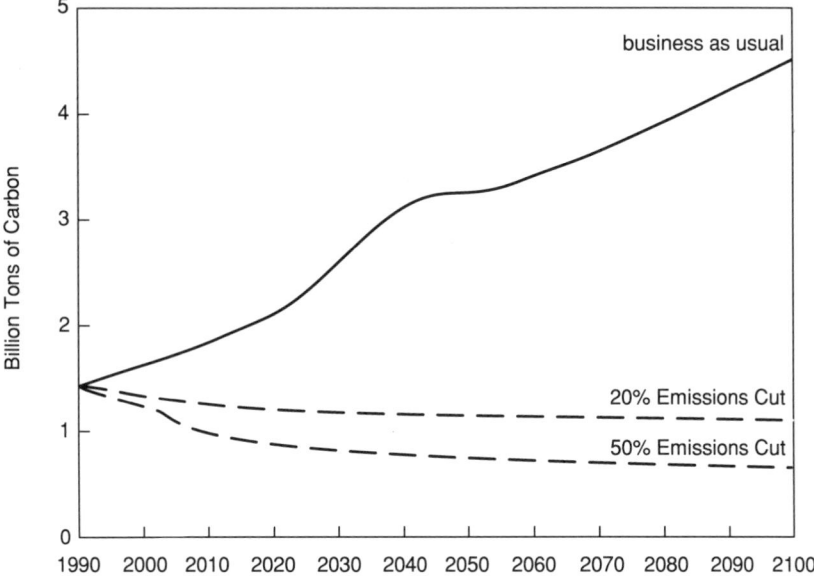

Figure 3.18 Alternative emissions targets

expensive to go from business as usual to 20 percent as it is to go from 20 percent to 50 percent. (For more details, see the willingness-to-pay discussion in chapter 6.)

Summary and Conclusions

CO$_2$ emission cuts are often justified from an insurance perspective. Activists argue that the risks of irreversible environmental damage are large, important scientific uncertainties will not be resolved for many years, and the costs of reducing CO$_2$ emissions are low. If emissions can indeed be significantly reduced at little or no cost, emission constraints make a good deal of sense. Immediate controls represent a reasonable hedge against unacceptably rapid climate change. If, on the other hand, the skeptics are correct and the size of this insurance premium turns out to be high, it may be worth pursuing other alternatives to immediate controls.

In this chapter, we have examined the costs of alternative carbon constraints to the U.S. economy. These cost estimates are quite sensitive to one's views about the future character of the energy system. One can point to plausible assumptions that will result in abatement costs on the order of several trillion dollars (discounting year-by-year losses back to 1990). Alternatively, if one is optimistic about the potential for supply and conservation technologies, the costs could be much lower.

The wide range of estimates presents a dilemma for decision makers. Do we impose sizable controls during the near future and run the risk of incurring large economic costs—with little or no beneficial effect? Or do we wait for greater scientific knowledge and run the risk of irreversible environmental damage? Policy makers must weigh the costs of premature action against those of delay. In the next chapter, we present a framework for analyzing the conflicting risks that are inherent in the greenhouse debate. We examine the near-term decisions facing policy makers, explore their sensitivity to the long-term uncertainties, and quantify the value of improved scientific information.

References

J. T. Houghton, G. J. Jenkins and J. J. Ephraums. 1990. *Climate Change—the IPCC Scientific Assessment*. Cambridge University Press, Cambridge.

T. J. Woods. 1988. "The Long-Term Trends in U.S. Gas Supply and Prices: The 1988 GRI Baseline Projection of U.S. Energy Supply and Demand to 2010." Gas Research Institute, Washington D.C., December.

4　Decision Making under Uncertainty

Introduction

The greenhouse debate is enlivened but not necessarily illuminated by rhetoric on "an irreversible ecological catastrophe" versus "the staggering costs of reducing emissions." Suppose that instead we take the view of an insurance purchaser who knows that the climate experts are deadlocked on the chances of a global calamity if we follow a business-as-usual policy. What steps should we take today to reduce the risks to future generations?

What are the options? Three forms of greenhouse insurance dominate current discussions: continued intensive science research to reduce climate and impact uncertainties, development of new supply and conservation technologies to reduce abatement costs, and immediate reductions in emissions in order to slow down climate changes. Although these options define the rough outlines of the debate, there is little agreement on their relative merit. Some argue that it is premature to think about doing anything other than intensive research. Others claim that scientific certainty on the greenhouse issue is an elusive goal and that the risks of waiting are simply too great.

Such perspectives tend to oversimplify the choices. The issue is not one of either-or but one of finding the right blend of options. Uncertainty need not lead to paralysis. Policymakers must decide how to divide greenhouse insurance dollars among competing needs. What portion goes to resolving climate uncertainties? What portion goes to technology development? And what portion goes to immediate abatement of emissions?

The key to selecting an optimal portfolio is to understand how the options interact. For example, better greenhouse information is likely to influence technology development and emission reduction decisions. Similarly, the prospect of low-cost, carbon-free supply alternatives and highly efficient end-use technologies will affect the need for better scientific information and the allocation of emission rights among generations.

In this chapter, we analyze some of these interactions. A logical starting point is the issue of near-term emission cuts. There have been a number of recent proposals to reduce emissions in the industrialized nations. Proponents argue that such reductions are necessary as a hedge against unacceptably large greenhouse warming. It is important to understand how the optimal hedging strategy varies with the accuracy and timing of climate research and with the prospects for new supply and conservation technologies.

We also look at the value of improved information. The United States and other IPCC (Intergovernmental Panel on Climate Change) participants have undertaken a substantial research program in order to resolve climate-related uncertainties. Ideally, improved information will lead to better decisions. What is the value of reducing scientific uncertainty? How much accuracy is needed in order to improve decisions on carbon limitations? How do the prospects for new supply and conservation technologies affect the value of information?

Global 2100 takes account of realistic time lags in adapting both supplies and demands to changes in the value of carbon emission rights. It allows for the role of price-induced and nonprice energy conservation. Uncertainty is a major factor in the greenhouse debate. We begin with a back-of-the-envelope calculation to introduce some of the basic concepts that underlie an analysis of this type and then provide a more rigorous treatment based on a probabilistic model of the United States.

The Value of Information: An Illustrative Calculation

Our first calculations are illustrative but far from definitive. We sketch out a decision-theoretic framework for handling the uncertainties inherent in the global greenhouse controversy. The basic idea is to avoid the difficulties inherent in the debate over whether the benefits of control outweigh their costs. Instead, the analysis is focused on identifying the value of information in making near-term decisions on emission reductions. In this first example, it is assumed that by 2020 a political consensus will emerge that the benefits of control are very large

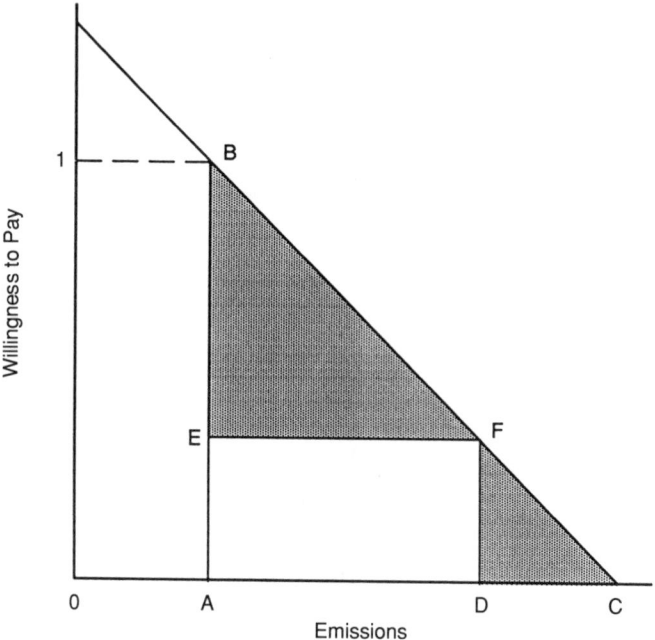

Figure 4.1 Willingness to pay for greenhouse emissions

or that they are negligible. Explicit probabilities are assigned to these two alternative long-term scenarios.

Consider the single-period demand curve for emission rights shown in figure 4.1. This represents the marginal value of emission rights, and the area below the demand curve provides a measure of the total cost of emission cutbacks. For simplicity, willingness to pay for carbon rights is shown as a linear function of the quantity of emissions. The identical function is used in order to characterize demands during the two thirty-year periods before and after the resolution of climate uncertainties in 2020.

In the absence of a greenhouse tax, emissions during each period would be OC. With a sufficiently high tax (AB), emissions could be reduced to OA. Since CO_2 has a long atmospheric residence time, all that matters is cumulative emissions. It does not matter whether cutbacks are imposed during the first or the second thirty-year period. Now suppose that the greenhouse pessimists define AC as the total cumulative amount of reductions required to avert a global catastrophe. One possibility might be to hedge by undertaking a small reduction (DC) during the first period and a large reduction (AD) during the second period—if the pessimists are indeed correct. By assumption, the greenhouse danger will be identified early enough so that there is no long-term irreversible damage from a hedging policy. In the worst case (the business-as-usual scenario with no reductions in the first period and DC = 0), all that is required is a rapid replacement of capital stocks (housing, power plants, and so forth) during the years 2020 through 2050.

Without loss of generality, we can choose the units of emissions and of willingness to pay so that the distances AC = AB = 1. The maximum losses are therefore .50, the area of the triangle ABC.

Let x be the decision variable that denotes the fraction of the total reductions to be undertaken during the first period, and let $(1 - x)$ represent the fraction of the reduction during the second period. That is, x corresponds to the distance DC, and the lower shaded triangle DCF corresponds to the immediate costs of this decision. The second period's emissions are AD = $(1-x)$. The upper left shaded triangle BEF corresponds to the second period's costs. The rectangle ADFE indicates the cost savings if we were certain that the cumulative cutback of AC will be required. The costs during the first and second period are, respectively $.5x^2$ and $.5(1 - x)^2$.

Let p be the probability of a political consensus that the greenhouse effect would be catastrophic and assume that this uncertainty is resolved by 2020. Without time discounting, the expected money value of the costs incurred during the two periods together would depend as follows upon the probability p and the near-term decision variable x : $EMV(x) = .5x^2 + .5p(1 - x)^2$.

With this cost function, it is straightforward to show that the optimal policy is to set $x = p/(1 + p)$. That is, if we were certain that the greenhouse phenomenon will prove to be catastrophic (p = 1), the optimal policy would be to undertake half the cumulative reduction during the first period, and half during the second. Similarly, if there is only a fifty-fifty chance of a catastrophe, the optimal hedging strategy would be to commit to a 33 percent reduction in the near term and a 67 percent reduction subsequently, should this prove to be necessary. This is a low-regrets strategy. In this example, there is no way to construct a no-regrets policy.

If p = .5, and therefore the optimal x = .33, EMV(.33) = .167. These are the expected costs associated with decisions under uncertainty ("act–then learn"). By contrast, under a scenario approach ("learn–then act"), all uncertainties are resolved prior to the date at which decisions are taken. The expected value of costs would then drop to .125. This can be seen as follows: (1) If we know that there is no need to act, the cost is zero in both time periods. (2) If we know that we do need to act, the optimal policy is to set $x = .5$. The total cost in the two periods will then be: $.5(.25 + .25) = .25$. Since both outcomes are equally likely, the expected costs are .125. Therefore the expected value of perfect information, EVPI = $.167 - .125 = .042$. Note that the EVPI = $.042/.167 = 25$ percent of the expected costs under an act–then learn policy. In plain English, this means that there is a large payoff to the resolution of uncertainties prior to 2020.

This example is suggestive, but it needs to be extended to include more realistic features. The following analysis includes nonlinear demand functions that shift from one period to the next. It includes time discounting and economic growth. Each of these ideas may be incorporated within the Global 2100 model, but we retain the basic ideas underlying the two-period example. That is, if there is a positive probability that carbon emissions will eventually have to be limited, how much greenhouse insurance should we buy during the near future? Is it worthwhile to delay the reduction of emissions until improved climate modeling information becomes available?

A Framework for Dealing with Uncertainty

In chapter 3, there was only an indirect treatment of uncertainty. For a given scenario, all uncertainties were resolved prior to decision making. The analysis proceeded as if we had the opportunity to learn the state of the world before taking action. Scenario analysis is one way to look at the here-and-now significance of alternative futures. Figure 4.2a shows the decision tree for this "learn–then act" characterization of the decision problem. A circle denotes a chance node—a point at which an uncertainty is resolved. A square denotes a decision node—a point at which actions are required. Figure 4.2b indicates the basic structure that underlies our introduction of uncertainty into Global 2100. It is characterized by the phrase "act–then learn." Decisions are taken at discrete points of time separated by intervals of one

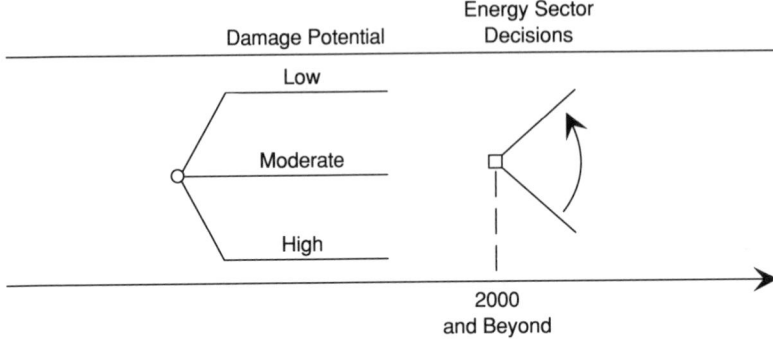

a) Learn, Then Act

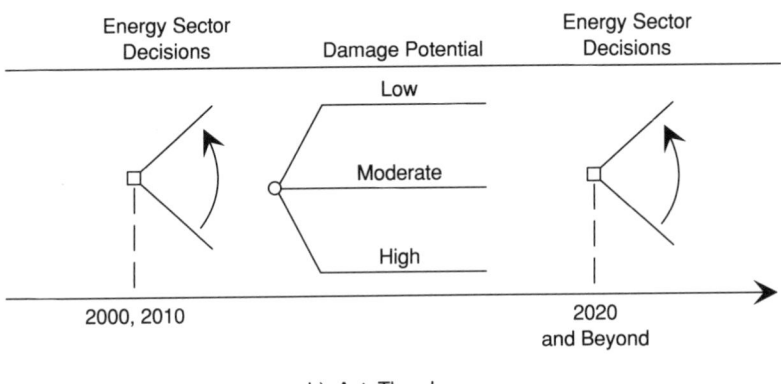

b) Act, Then Learn

Figure 4.2 Alternative characterizations of decision problem

decade. Prior to 2020, the energy sector's supply and demand decisions must be made under uncertainty about reaching a consensus on carbon limits.

From 2020 onward, decisions will be made after the resolution of uncertainty. In order to introduce uncertainty into the model, a state-of-world (SOW) subscript is attached to each of the original decision variables and constraints. During the initial periods (2000 and 2010), the decision variables must take on identical values for all three SOW, but during the subsequent periods (from 2020 on), they may take on different values, depending on the magnitude of the carbon constraint. (For further technical details, see box 7.3.)

Table 4.1
Emissions level

Damage potential	Carbon emissions
Low	Unlimited
Moderate	20% below 1990 level
High	50% below 1990 level

Just as in the two-stage model suggested by figure 4.1, the cumulative allowable carbon limit is determined exogenously by a probabilistic mechanism. However, the period-by-period allocation of carbon emissions is determined endogenously so as to maximize the expected discounted utility of consumption.

In order to introduce uncertainty into Global 2100, two additional sets of parameters are needed: alternative carbon emissions scenarios and the probabilities of realization of these scenarios. Figure 4.2 shows three long-term possibilities regarding the damage potential associated with the continued buildup of greenhouse gases. For each of these possible outcomes, there is an appropriate constraint on carbon emissions. The more dire is the outcome, the more drastic becomes the cutback in emissions.

Table 4.1 shows the emissions constraint ascribed to each of the three states of the world under a learn then act procedure. If the damage potential is low, carbon emissions remain unconstrained. Otherwise, emissions are held at their 1990 level through the turn of the century and then reduced to the desired target by 2010. With a moderate damage potential, the 2010 target is 20 percent below 1990 levels. With a high damage potential, the 2010 target becomes a 50 percent reduction.

Under act–then learn, we make the decisions for 2000 and 2010 without knowing which of these three states will occur and must select an optimal hedging policy for the emissions levels. This policy will depend on the probabilities of the alternative outcomes. To illustrate a case in which there is a high EVPI, suppose that the probability of low damage is 60 percent and that the probability of a 20 percent reduction is 1.5 times that of a 50 percent reduction. That is, if controls are required, they are more likely to be moderate than high.

With these probabilities, figure 4.3 provides a comparison between the optimal carbon emissions paths under learn–then act and act–then learn. Along the dashed lines, we make decisions for 2000 and 2010

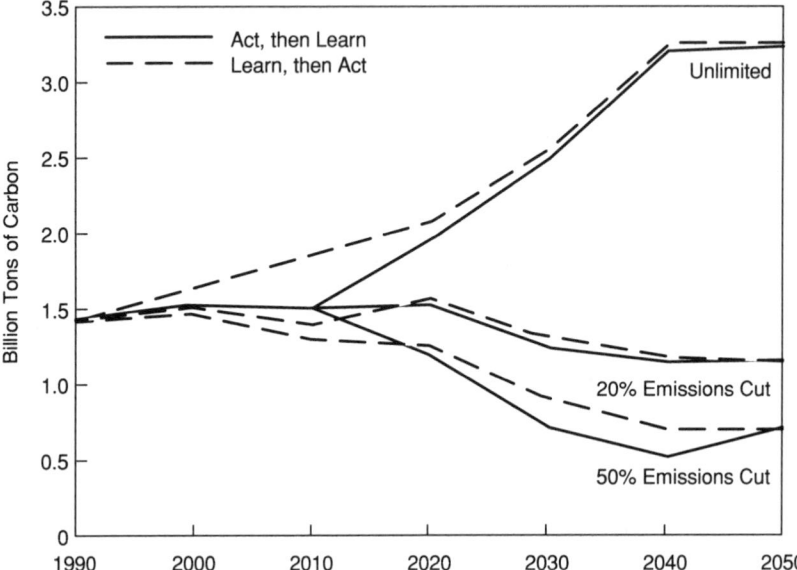

Figure 4.3 Carbon emissions

in full knowledge of whatever turns out to be the potential damages associated with global warming. Either the emissions are unlimited, or we begin to impose carbon constraints. The three paths diverge immediately after the 1990 base year. By contrast, the solid lines show what happens when the damage potential remains unknown during 2000 and 2010. In this case, the carbon emissions paths do not diverge until after 2010. The hedging strategy consists of adopting an emissions level that lies between the extreme cases shown along the dashed lines. Note that the optimal initial decisions (the emissions in 2000 and 2010) are different between the dashed and solid lines. There are losses associated with uncertainty. The probabilistic version of Global 2100 identifies a strategy that minimizes the expected value of these losses. Clearly the size of the hedging strategy will depend on the probabilities assigned to alternative scenarios.

In the two learn–then act scenarios for which emission cuts must be undertaken, the dashed lines lie below the solid line ("act–then learn") during the two initial periods: 2000 and 2010. Thereafter, they lie above. The solid lines eventually converge toward the dashed lines. This is consistent with the requirement that for each state of the world

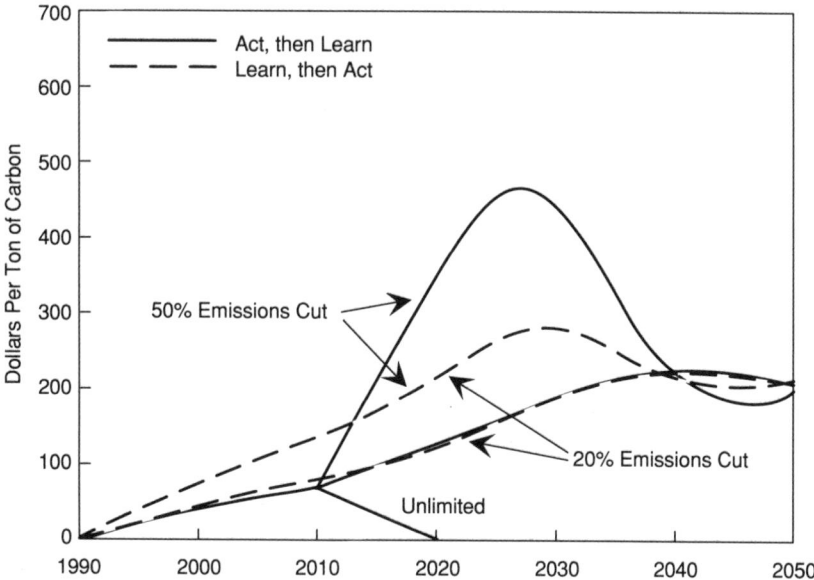

Figure 4.4 Carbon taxes

in which cuts are required, cumulative CO_2 emissions are identical over the long run.

An appropriate carbon tax is associated with each emissions scenario (figure 4.4). This is sometimes known as state-contingent pricing. With learn–then act, it is possible to adapt smoothly to whichever of the three states of world eventually occurs. With act–then learn, we do not know the state of the world during the initial years. The tax rate must be identical until the uncertainties are resolved. There is an abrupt transition at the point when we learn which of the three carbon limitations must be adopted. If a 50% carbon limitation is required, there are heavy costs of a rapid transition and a substantial overshoot above the backstop level of $208 per ton.

Perfect Information: A Special Case

A substantial research program is underway to reduce climate-related uncertainties. What is the value of such a program? First, consider the case of perfect information on a political consensus over carbon limitations. All uncertainties are resolved prior to the energy sector's decisions for the year 2000. Although this is straightforward from an analytical perspective, it is also quite unrealistic. Nonetheless, it represents a useful point of departure, placing an upper bound on the value of a climate research program. In figure 4.5, the value of perfect information is shown as a function of the subjective probability that emissions will remain unconstrained. (Throughout the following computations, the probability of a 20 percent reduction is 1.5 times that of a 50 percent cutback, regardless of the probability that emissions will remain unconstrained.)

The value of perfect information is the value of being able to chart the right course from the outset. In the face of uncertainty, it is prudent to hedge our bets. Act–then learn leads to initial decisions that lie between the extreme values shown under learn–then act. With a hedging strategy, we will always be somewhat off course. Near-term actions are affected by the distant-future uncertainties. If the damage potential turns out to be low, we will have overly constrained carbon emissions during the near term. If the true damage potential turns out to be moderate or high, near-term emissions will have been excessive. In either case, getting back onto the long-term track will involve costs. These costs may be eliminated by knowing the true damage potential in advance.

According to figure 4.5, perfect information may be worth tens of billions of dollars in terms of its impact on macroeconomic consumption. Discounted to 1990 at 5 percent, the EVPI (expected value of perfect information) is at its maximum ($81 billion) when the probability of unlimited emissions is .60. The EVPI falls to zero as the probability approaches one. Note, however, that the curve is not symmetric. A zero probability of unlimited emissions does not imply a zero value of information. We would still not know whether a 20 or a 50 percent emissions cut will be required.

The greater is the uncertainty on the need for an emissions constraint, the higher becomes the value of information. On the other hand, if the true outcome is already known, further information will be of little value; it will only confirm that which is already believed.

The Value of Improved Information

Perfect information is an elusive goal. What is the value of improved information? In order to improve decisions on carbon limitations, how

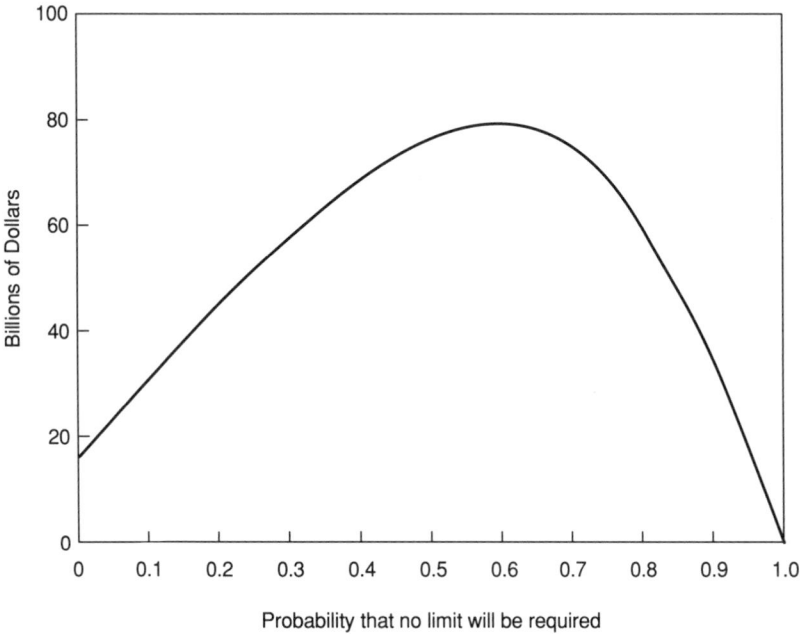

Figure 4.5 Expected value of perfect information

much accuracy is needed in climate modeling and in impact assessment? What if uncertainties are resolved in stages spanning several decades?

Figure 4.6 shows how the original decision tree can be modified to address these questions. Our initial decisions relate to how much we are willing to pay to reduce uncertainties about the damage potential. Assume that the research program will yield improved information soon enough to revise supply and conservation decisions for the year 2000. Regardless of the magnitude of research expenditures, it is unlikely that uncertainty can be completely resolved by 2000. Accordingly, we assume that the damage potential will remain unknown until well into the second decade of the twenty-first century. Only then will we learn the appropriate emissions level.

The decision tree structure reflects the fact that the energy sector's near-term supply and conservation decisions must be made in the face of uncertainty regarding the ultimate damage potential. The degree of uncertainty can be represented by a likelihood table indicating the probability that the damage forecast will be low, moderate, or high— given that the true damage potential is low, moderate, or high. This

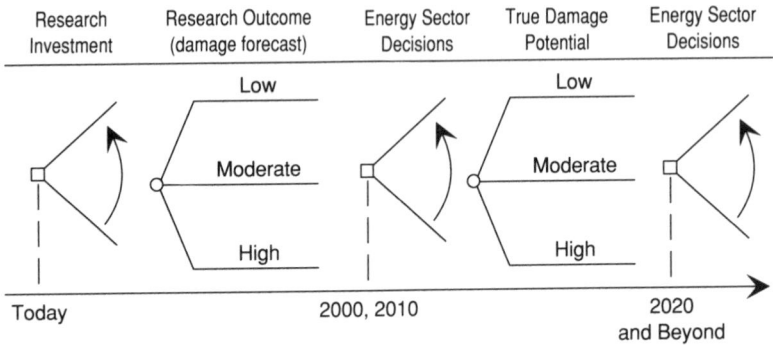

Figure 4.6 Decision tree for value of information analysis

representation allows the use of standard probabilistic techniques to incorporate climate research outcomes into the carbon decision model.

Table 4.2 illustrates the conditional probabilities for the research outcomes. The information would be perfect if the diagonal entries in the table were equal to unity and all other cells were zero. In this case, there could be no error associated with any forecast resulting from the climate research program. What if the research program turned out to be useless, and the forecast were independent of the actual damage potential? This case would be represented by equal entries in all cells of the likelihood table.

A reasonable research program should produce something less than perfect and something better than no information. The accuracy of the information depends on the particular characteristics of the likelihood table. In table 4.3, we assume that the errors are symmetrically distributed. If a forecast is wrong, it is equally likely to be wrong on the low side as on the high side. The parameter σ is a measure of the accuracy of the forecast.

Consider two extreme cases. If $\sigma = 1/3$, the research program is useless. All forecasts would be equally probable regardless of the damage potential. Conversely, if $\sigma = 1.0$, this would be the case of perfect information. Accordingly, the accuracy of climate forecasts will increase with σ over the interval between 1/3 and 1. We now show how Global 2100 may be employed to estimate the economic returns from a research program with different values of σ.

Table 4.2
Conditional damage forecasts

Research outcome (damage forecast)	True damage potential		
	Low	Moderate	High
Low	Pr (low forecast/ low damage	Pr (low forecast/ moderate damage	Pr (low forecast/ high damage
Moderate	Pr (moderate forecast/ low damage	Pr (moderate forecast/ moderate damage	Pr (moderate forecast/ high damage
High	Pr (high forecast/ low damage)	Pr (high forecast/ moderate damage)	Pr (high forecast/ high damage)

Table 4.3
Assumed likelihood matrix

Research outcome (damage forecast)	True damage potential		
	Low	Moderate	High
Low	σ	$(1-\sigma)/2$	$(1-\sigma)/2$
Moderate	$(1-\sigma)/2$	σ	$(1-\sigma)/2$
High	$(1-\sigma)/2$	$(1-\sigma)/2$	σ

Hedging Strategies under Imperfect Information

Imperfect information is represented as a case in which the accuracy parameter is less than unity. When $\sigma = 1$, the climate research program will unequivocally reveal the true damage potential soon enough so that we can revise supply and conservation decisions for the year 2000. When $\sigma < 1$, the climate research results will be imperfect. Just as in weather forecasting, there will be a margin for error.

Now continue to assume that the probability of a low damage potential is .60. Figure 4.7 illustrates the relationship between information value and information accuracy. Note how quickly the value of information drops off as we adopt lower values for σ. For example, if there are four chances in five that the research results are accurate ($\sigma = .80$), the value of information falls to $35 billion. This is still significant but less than half the value of perfect information. If there are only three chances in five that the results are correct ($\sigma = .60$), the value drops to approximately $10 billion. In other words, it is worth $25 billion to increase research accuracy from .60 to .80.

The benefits from a research program are closely tied to our confidence in the results. Perfect information eliminates the need to hedge against surprises. The more faith we have in the forecast, the less we need to hedge. The top part of figure 4.8 shows the optimal hedging policy when the research program provides no new information ($\sigma = .33$). The dashed lines display the paths under learn–then act. These are the policies that would be chosen if we knew the true damage potential in advance. By contrast, the solid line indicates the optimal hedge that is associated with act–then learn.

To understand the reasoning that underlies figure 4.8, consider a specific example. Suppose that before committing to an emissions strategy we are able to observe the results of a research program, and

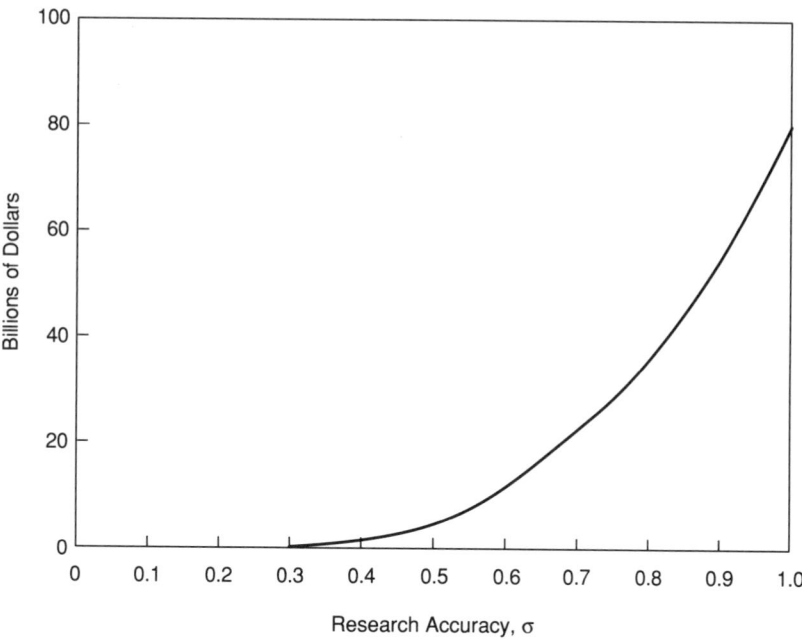

Figure 4.7 Expected value of imperfect information. Note: Research indicates low damage potential.

that our confidence in the findings is expressed by setting $\sigma = .80$. Further, suppose that the research indicates a low value of the potential damage. Figure 4.9 is a standard application of Bayesian analysis. It shows how the research outcome leads to a revision of our beliefs. Prior to receiving the information, the odds of a low damage potential are six chances in ten. They now rise to more than nine in ten.

The bottom part of figure 4.8 shows the impact on the optimal hedging policy. As one would expect, if the results point to a low damage potential, we move closer to the unlimited emissions strategy. The research program has strengthened our belief that the true damage potential is low and thus reduced the need to hedge against large future emission cuts.

Better information need not result in business as usual—that is, in an unconstrained emissions policy. If the research results indicate a significant damage potential, rational decision makers would modify their beliefs accordingly and adopt a more restrictive emissions policy. The value of a climate research program stems from its potential

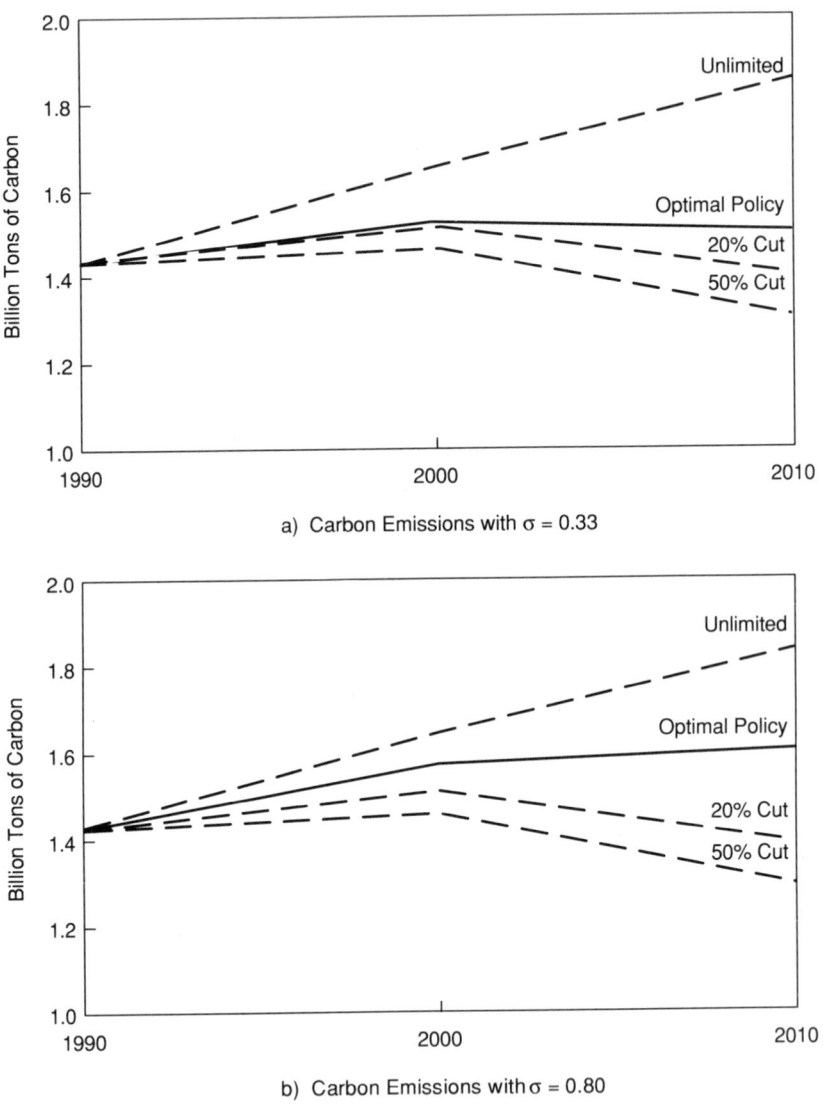

a) Carbon Emissions with σ = 0.33

b) Carbon Emissions with σ = 0.80

Figure 4.8 Optimal policy as a function of σ. Note: Research indicates low damage potential.

ectives

ns

proposals have been made for immediate cutbacks in CO_2
Proponents argue that reductions are necessary as a hedge
acceptably rapid changes in climate. This chapter examined
ptimal strategy might vary with the probability of eventual
mitations, the accuracy and timing of the climate research
and the prospects for new supply and conservation tech-

alysis yields several insights. We find that emission cutbacks
sensitive to the quality and timing of the climate research
n the best of all possible worlds, uncertainties would be com-
resolved prior to energy sector decision making. With perfect
tion today, we could chart the best course of immediate action.
vould be no need to hedge bets. By contrast, the less faith we
the possibility of near-term research accuracy, the greater is
d for precautionary actions.
straightforward to calculate the benefits from reducing scientific
ainty. Better information leads to better decisions. Global 2100
us to calculate the aggregate economic impacts. The analysis
little doubt that there can be a big payoff to reducing climate-
d uncertainties and that it could be upward of $100 billion for
nited States alone. At the same time, we find that the value
formation is quite sensitive to confidence in its timeliness and
 bility. Relatively small increases in accuracy can yield substantial
ends.
e cutback strategy is sensitive to the prospects for new supply and
ervation technologies. Consider two alternative energy futures.
he optimistic scenario, there are abundant low-cost, carbon-free
ly alternatives and highly efficient end-use technologies. In the
simistic scenario, there is heavy dependence on carbon-intensive
thetic fuels and less autonomous conservation. Clearly, with the
er scenario, it is more difficult to adapt to a carbon constraint. If
e is a technology pessimist, it may make sense to shift some of
burden of emissions abatement from the future to the present
neration. For the optimist, the story is reversed: future generation
ll need little help in adapting to a carbon constraint, so there is les
essure for immediate cutbacks.

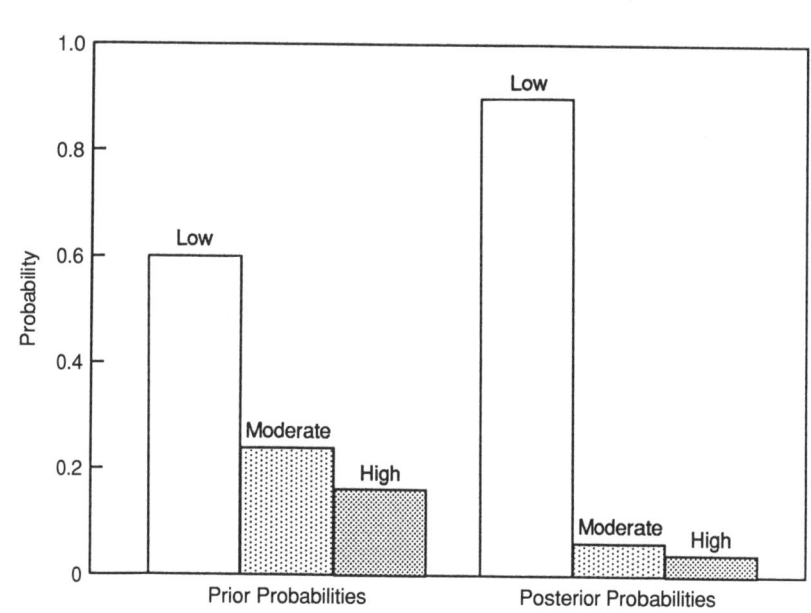

Figure 4.9 Effect of low damage forecast on subjective probabilities.

to modify near-term decisions. The more confidence we attach to its
results, the less we need to hedge.

The Impact of Supply and Conservation Estimates

The value of information is sensitive to one's degree of optimism
or pessimism concerning new technologies. This can be illustrated
through the alternative supply and demand scenarios described in
chapter 3. Recall that the optimistic scenario describes an energy future
in which low-cost, carbon-free supply alternatives are in greater supply and there is a lower overall demand for energy. In the pessimistic
scenario, there are fewer options for energy supplies and conservation.

The top three solid lines in figure 4.10 show the implications for
carbon emissions when no limits are imposed. Under the optimistic
scenario, emissions rise more slowly than in the base case. There are
low-cost, carbon-free substitutes for coal-fired electricity and crude oil,
and there are lower energy demands. Conversely, for the pessimistic
scenario, there is greater dependence on coal in both the electric and

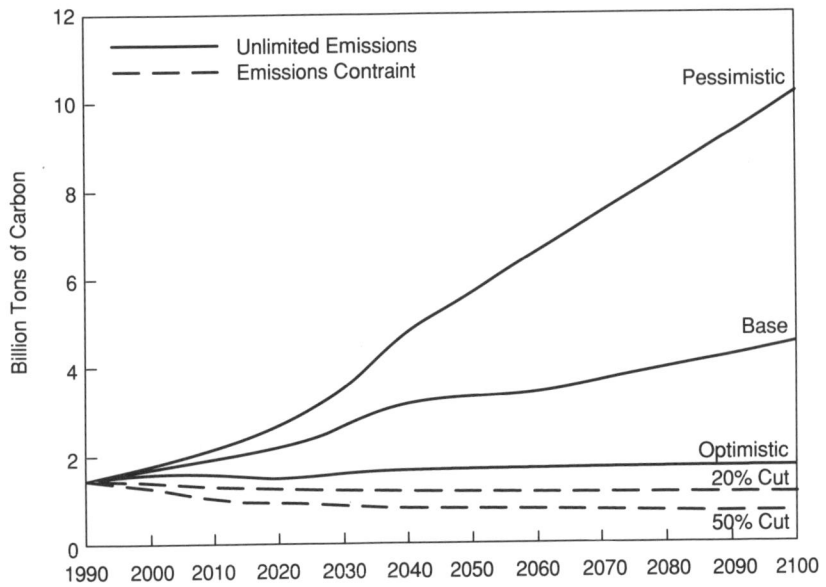

Figure 4.10 Carbon emissions for three supply/demand scenarios

nonelectric sectors, there are higher demands for primary energy, and there are higher carbon emissions.

These differing perspectives have a direct impact on the EVPI (figure 4.11). Technological optimism lowers the value of information. Why? Figure 4.10 shows that optimism leads to low carbon emissions even under a business-as-usual scenario. There is a smaller difference between the constrained and unconstrained paths than in the base case. When the constrained and unconstrained scenarios lie close together, there are lower costs associated with being off course. This reduces the value of knowing the desired path in advance. Conversely, when the constrained and unconstrained scenarios lie farther apart, there are higher costs of being off the right track. Technological pessimism tends to raise the value of information.

According to figure 4.12 (based on act–then learn), technological optimism reduces the size of the optimal hedge. In this case, the constrained and unconstrained emission paths lie close together. Future generations will have less difficulty in adapting to a carbon limit. This considerably reduces the incentive for immediate abatement measures. By contrast, technological pessimism increases the hedge size.

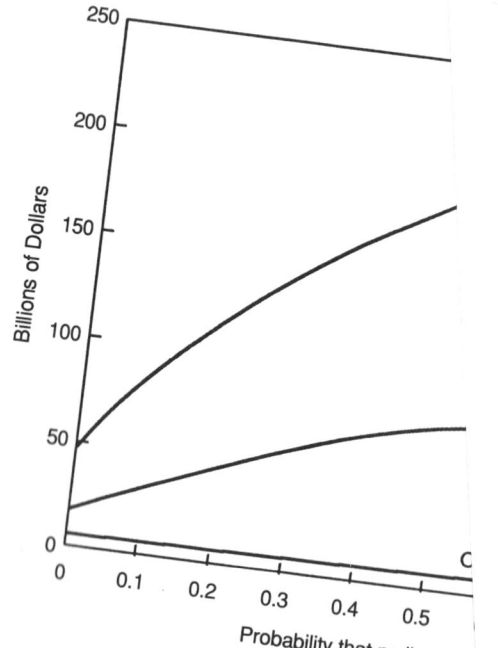

Figure 4.11 EVPI for three supply/demand scenario

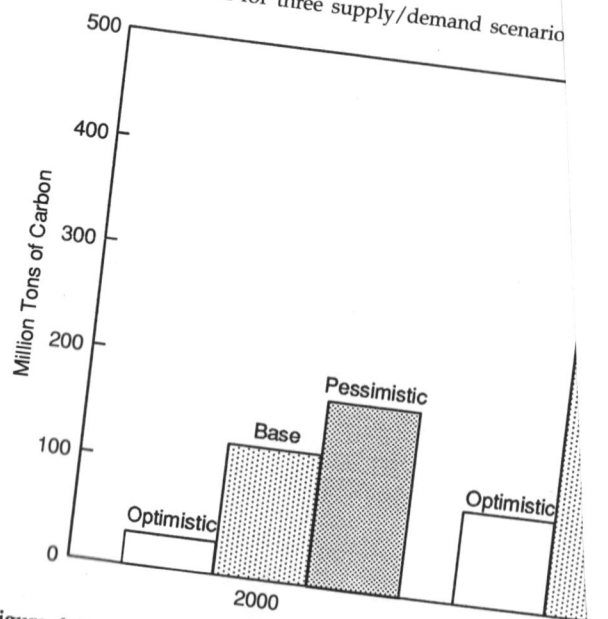

Figure 4.12 Optimal emissions reductions for three supply/der
Note: Reductions are relative to unconstrained cases.

Conclusio

Numerous
emissions.
against un
how the
carbon li
program,
nologies.

Our an
are quite
results.
pletely
informa
There
have i
the nee

It is
uncert
allows
leaves
relate
the U
of in
reliab
divic

Th
cons
In t
sup
pes
syr
lat
on
th
ge
w
p

The near-term policy implications are clear. There is less need for precautionary emission reductions if we undertake a sustained commitment to reducing climate uncertainty and developing new supply and conservation options. Better climate information reduces the need to hedge against a potentially hostile future. Improved supply and conservation technologies will improve our ability to deal with such a future should it occur.

5   A Global Cost Analysis

Introduction

The greenhouse effect is inherently a global issue. If significant reductions in emissions are required, they can be accomplished only through international accords. Negotiations are bound to be difficult because of enormous disagreements over how far greenhouse gas emissions should be reduced, how reductions should be allocated, and who should pay (Grubb 1989).

Although calls for reducing CO_2 emissions have been a common theme at international conferences on global warming, the proposals vary widely. They range from slowing the rate of growth to halving the absolute level of current emissions. Rather than arbitrary percentage targets, a sensible policy will involve balancing incremental impacts and costs (National Academy of Sciences 1991). This will be difficult to do without a better understanding of the consequences of global warming. More detailed study is needed on the environmental, economic, and social impacts of the continued accumulation of CO_2 in the atmosphere.

Even if we knew the optimal emissions level, we would still face daunting questions of how reductions should be allocated and who should pay. For a set of limits to be broadly acceptable, they must be perceived as equitable. Unfortunately, there is no unique definition of fairness. Should industrialized countries be penalized for their past emissions? How should income and population growth be factored into the equation? Is the ultimate goal equal per capita emissions? The burden-sharing issue is likely to be a major impediment to an international agreement.

This book does not attempt to determine global targets for emissions reduction, nor do we put forward a specific proposal for an equitable distribution of emission rights. We do, however, attempt to provide essential information for these decisions. The choice of an optimal emissions level requires analysis of both benefits and costs. Our goal is to provide insights into the latter. Potential signatories to any agreement will want to know what sacrifices are entailed for each participant in a particular burden-sharing scheme. We provide an explicit analysis of how costs might be distributed among regions.

We continue to rely on Global 2100. The globe is disaggregated into five major geopolitical groupings: the United States, other OECD nations (Western Europe, Canada, Japan, Australia, and New Zealand), the former Soviet Union, China, and the rest of the world (ROW). Although the identical logical structure is employed throughout the world, there are major differences in the numerical values that characterize each of the five regions. The regions differ, for example, in terms of their projected GDP growth rates, their prospects for price- and nonprice conservation, and their resource endowments.

Suppose, for example, that we follow the conventional wisdom of petroleum geologists (Masters et al. 1987). At 1990 rates of consumption, the United States would exhaust its natural gas reserves and undiscovered resources in less than 50 years. The Soviet Union's natural gas would last for nearly 170 years. We cannot expect that both countries will take the same time to complete their transition away from exhaustible hydrocarbons. By contrast, China is poorly endowed with oil and natural gas but has large deposits of coal. This fact alone should alert us to the prospect that China is likely to suffer large costs if it is required to make drastic emissions cutbacks.

A Business-as-Usual Scenario

Globally, carbon emissions increased at an average annual rate of 3.3 percent per year between 1950 and 1990. Over this period, there has been a substantial shift in the pattern of global contributions. In 1950, the industrialized nations (the United States, other OECD, and the Soviet Union) were responsible for 9 of every 10 tons of carbon emitted into the atmosphere resulting from the combustion of fossil fuels. By 1990, their share had dropped to 64 percent. By contrast, the portion attributable to China and India rose from a negligible amount to nearly 15 percent.

In order to estimate the cost of CO_2 emissions abatement, we begin with an estimate of the size and pattern of future emissions under

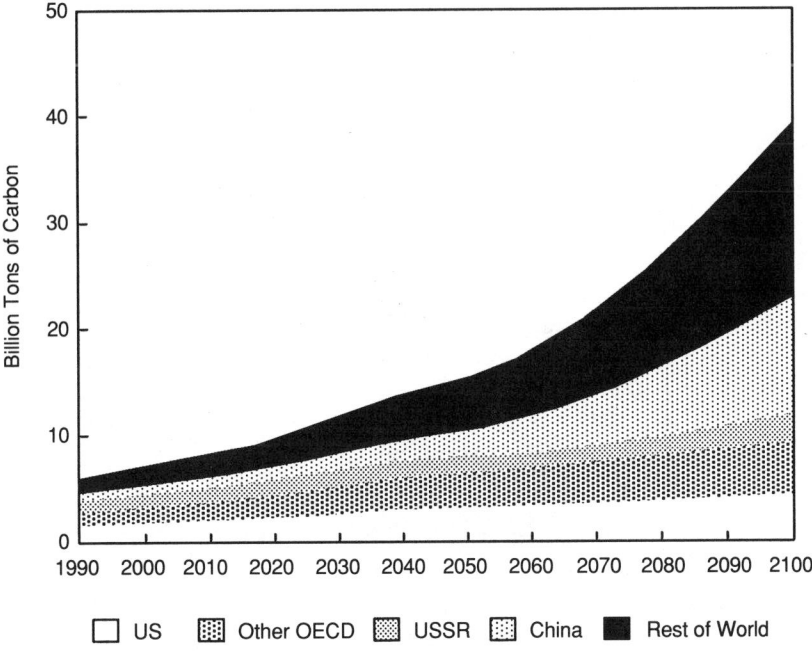

Figure 5.1 Carbon emissions, business as usual

an unconstrained business-as-usual energy future—that is, with no attempt to limit the growth of CO_2. Using the base case supply and demand parameters, figure 5.1 shows how the situation might evolve under this scenario.

Between 1990 and 2100, global emissions grow by a factor of nearly 7. Although this seems large, the average annual growth rate is only 1.7 percent, well below the rate projected for global GDP. In order to understand the overall trend in carbon emissions, one must allow for both the direct and indirect effects of the gradual exhaustion of conventional oil and gas during the twenty-first century. With rising oil prices, there is a stimulus for price-induced energy conservation. On the other hand, rising oil prices provide a stimulus for the introduction of coal- and shale-based synthetic fuels. Synthetic fuels have a carbon coefficient that is about twice that of conventional oil. In terms of their impact on carbon emissions, synthetic fuels could offset many of the gains from price-induced energy conservation.

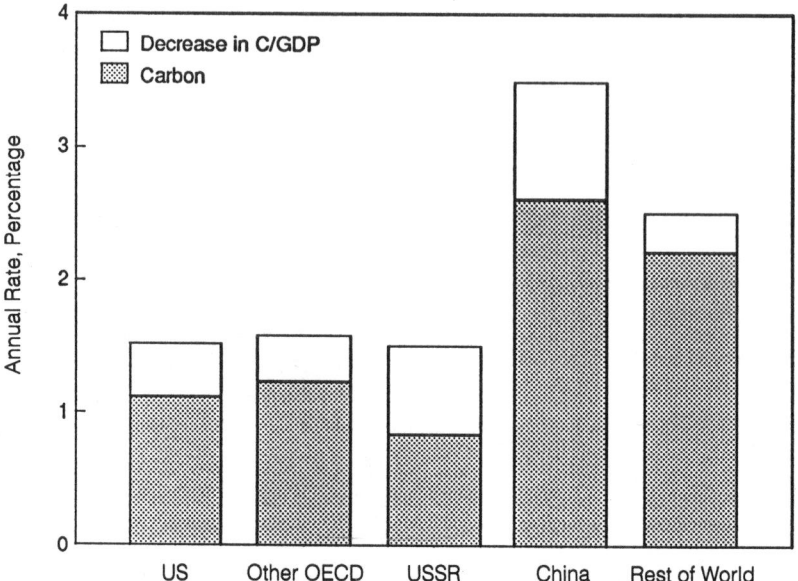

Figure 5.2 Carbon and GDP growth rates for five regions, 1990–2100. Note: Components add up to rate of GDP growth

According to our base case, there will be a significant shift in the pattern of global contributions to CO_2 emissions. In 1990, the industrialized nations accounted for 64 percent of the total. By 2100, their contribution is projected to drop to 30 percent.

Figure 5.2 provides useful insights into the regional distribution of changes. GDP growth rates are considerably higher in the developing countries. They therefore have higher growth rates in energy consumption and in carbon emissions. Notice that there is some decoupling between growth in GDP and growth in carbon emissions. The degree of decoupling varies between regions. For example, because of its vast natural gas resources, the Soviet Union is able to decrease carbon emissions at a faster rate per unit of output than its Western counterparts. The same is true for China, but here the reason is the high value assumed for autonomous energy efficiency improvements. Despite this optimistic assumption with respect to its AEEI, China's carbon emissions grow at the fastest rate among our five regions.

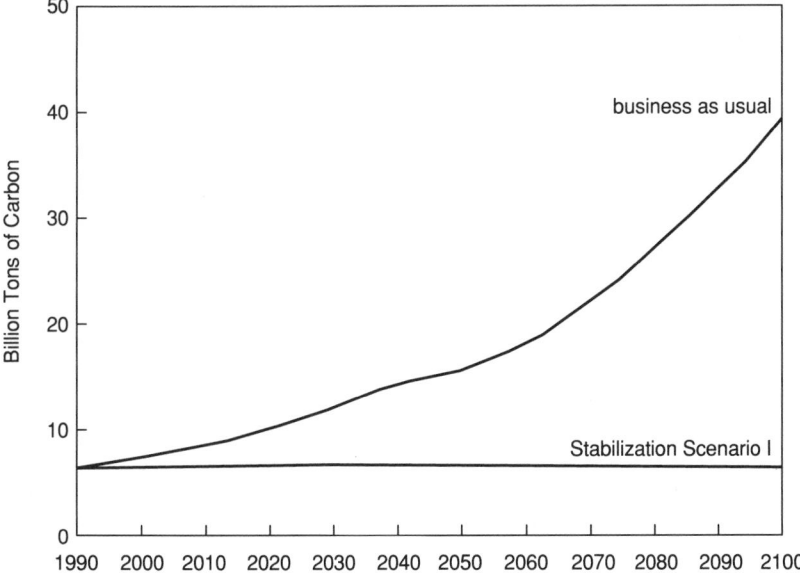

Figure 5.3　Global carbon emissions (with and without limits)

A Stabilization Scenario

With business as usual, there could be a major impact on climate at some point during the twenty-first century (Houghton et al. 1990). Accordingly, let us suppose that international negotiators reach agreement on the basic principle of stabilizing global emissions at 1990 levels. Although this target is not as ambitious as has been advocated by some parties in the greenhouse debate, it nevertheless represents a drastic change from the business-as-usual view of the future.

One widely discussed proposal is that the industrialized nations agree to a 20 percent cut. Specifically, we assume that these countries agree to hold emissions at 1990 levels through 2000, and then reduce them 20 percent by 2010. For global stabilization, the developing nations would have to accept a 50 percent limit on their emissions growth. Since we will explore other allocation schemes, this is named stabilization scenario I. Figure 5.3 shows that by 2100, this represents an 86 percent reduction relative to business as usual.

No doubt each group of negotiators will come up with compelling reasons why its nation deserves larger shares of the global emissions total. One delegation could urge that its country is making a major contribution by reducing greenhouse gases other than CO_2. Another could point out that its 1990 carbon-GDP ratio is already much lower than the global average. A third delegation could note that its people are poor and cannot afford to bear the burden of protecting the global environment. Scenario I will not satisfy all parties, but it is typical of the compromises that would have to be struck if a realistic international CO_2 agreement were to be reached.

The Costs of Stabilizing CO_2 Emissions: No International Trade in Carbon Rights

The feasibility of any scheme to reduce worldwide carbon emissions will depend on the costs to individual nations. Using Global 2100, we can add together the impacts of rising energy costs in each region and calculate the annual losses due to the carbon constraint. We begin by assuming that there is no international trade in carbon rights.

Figure 5.4 shows the annual losses as a percentage of conventionally measured GDP. For the United States, the losses begin immediately, and they keep rising until 2030. By that date, about 2.5 percent of the GDP is lost as a consequence of the carbon constraint. These percentages remain roughly constant for the remainder of the time horizon.

In the other OECD countries, the annual losses are limited to a range of 1 or 2 percent. We estimate that the costs of carbon constraints would be somewhat lower in that region than in the United States. They have a relatively high proportion of undiscovered oil and gas resources, and their nuclear power industry is larger. Perhaps more important, they start off with a lower initial energy/GDP ratio (figure 5.5). This reflects a much greater degree of urbanization and reliance on public transportation. These factors are likely to persist into the twenty-first century and have a significant effect on percentage GDP losses.

The Soviet Union will find it more difficult to adjust to a uniform percentage emissions cutback than its western counterparts, at least through the early decades of the next century. The higher costs are the result of several factors. The Soviet Union uses more energy per unit of output than the United States or other OECD (figure 5.5). We also assume that it will have more difficulty in decoupling GDP and energy growth. (Recall that both the elasticity of price-induced substitution and the rate of autonomous energy efficiency improvements are lower for the Soviet Union than for the United States and other OECD.)

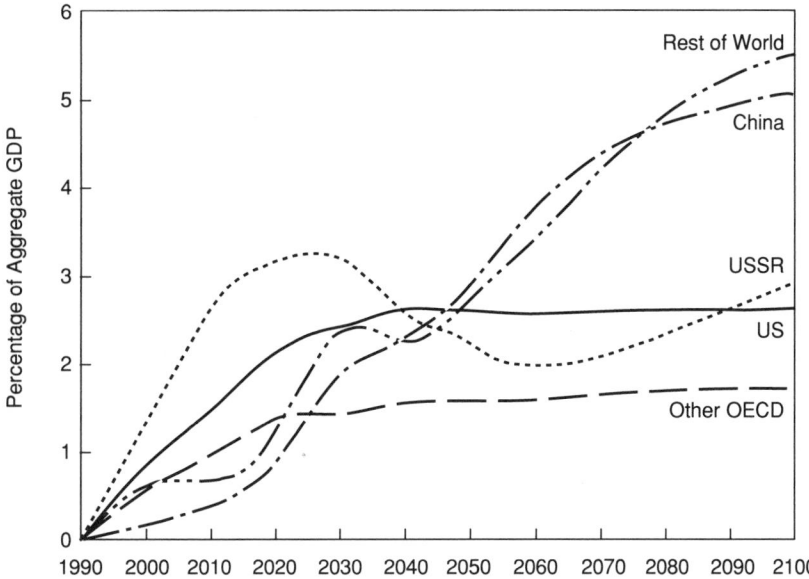

Figure 5.4   Annual losses under stabilization scenario I

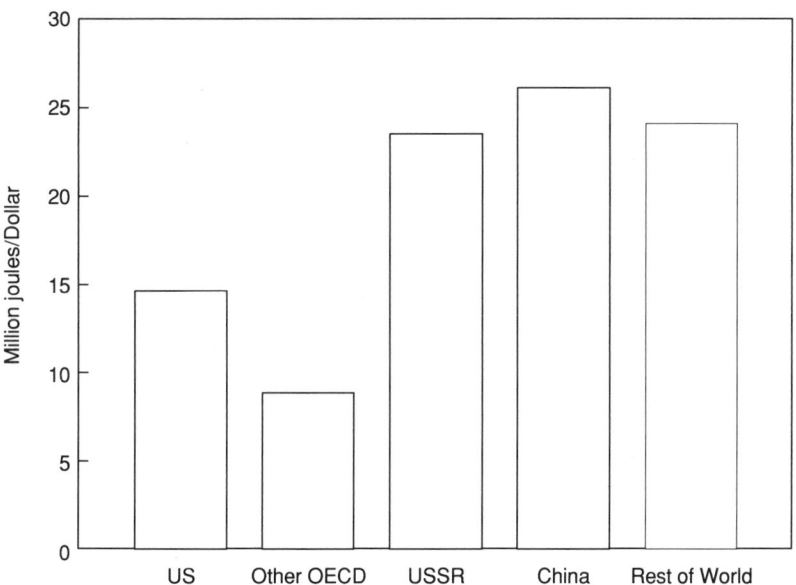

Figure 5.5   Total primary commercial energy/GDP ratios, 1990

This means that the Soviet economy is likely to remain more energy intensive than the OECD countries. It is not surprising that a penalty on the use of fossil fuels exacts a higher toll.

Despite the 50 percent increase allowed in their carbon emissions, it is costly for China and the ROW to accept stabilization scenario I. Their rapid rates of economic development generate enormous demands for commercial energy. Initially, their percentage GDP losses would be low, but eventually the developing countries become the regions most heavily affected by this proposal for an international carbon reduction agreement.

Carbon Taxes and the Value of Emission Rights: No International Trade

Figure 5.6 compares the time paths for the price signals required in order to implement stabilization scenario I. Before considering the case of international trade, we describe what might occur in its absence. In all regions but the Soviet Union, it turns out that the long-run equilibrium tax level is determined so as to make synthetic fuels (SYNF) and nonelectric backstop (NE-BAK) supply technologies equally attractive to energy consumers. Specifically, the equilibrium tax is governed by the ratio of the difference in cost to the difference in carbon content of these two technologies.

Despite the variations from one region to another, there is a general pattern to the trajectory of carbon prices. During the transition period, they tend to rise. Then, because of constraints on the rate of market penetration, they overshoot the backstop level. This pattern depends upon a specific feature built into the Global 2100 model: that it is permissible to delay the use of carbon rights. During the early years of the transition period, there are a number of cases in which it is optimal to exercise this option for delay. Accordingly, the value of carbon rises at the same rate as the marginal productivity of capital—about 5 percent annually in terms of dollars of constant purchasing power.

Under stabilization scenario I, all three industrialized regions would place a higher value on carbon rights than the developing countries during the early twenty-first century. The industrialized regions have agreed to reduce emissions during that period, and they lack sufficient supply- and demand-side alternatives to achieve such reductions without a sizable tax. By contrast, the developing regions are permitted a

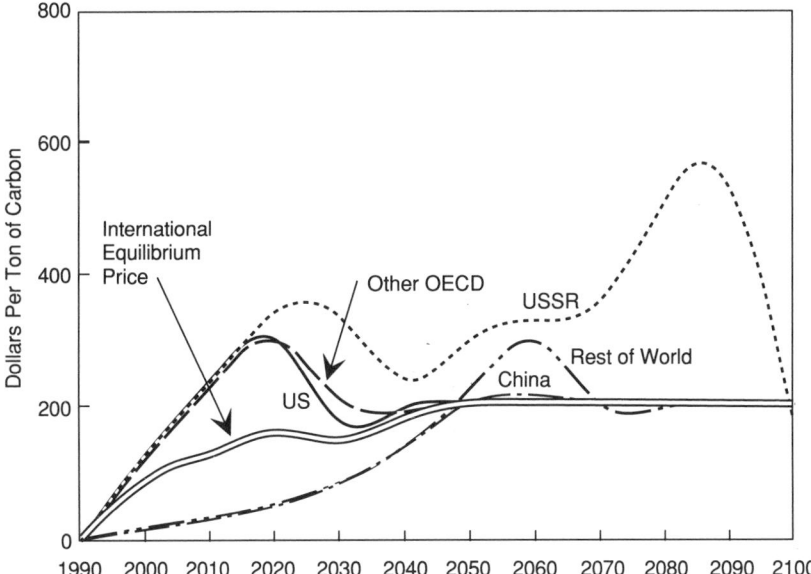

Figure 5.6 Carbon taxes, with and without trade

modest growth in emissions. During the first decades of the twenty-first century, the developing countries would find it attractive to sell some of their emission rights to the industrialized nations.

There are large interregional differences in the value of carbon, and these may be exploited to identify cost-effective strategies for international trade in carbon rights. At a given point in time, regions that find it more difficult to adjust to their emissions limits (those requiring higher taxes) should be willing to purchase emissions rights from regions experiencing less difficulty.

The Gains from Trade in Carbon Emission Rights

In our analysis of international trade, each individual region is viewed as a price taker and as a possible importer or exporter. Each is coupled to the others through the international price of carbon rights. In this type of projection, we believe that it is important to allow for consistent forward-looking expectations. Accordingly, the time path of prices must be determined so as to equilibrate supplies and demands during each period simultaneously. We no longer formulate Global 2100 as five independent optimizations. Instead, we solve an equilibrium model in which all choices are integrated through an international market in carbon rights. For this purpose, we have devised a special-purpose method of computations. (For technical details on the procedure, see chapter 8.)

Figure 5.6 provides important insights. During the backstop phase, it takes no more than a back-of-the-envelope calculation to determine the carbon tax. If all regions have identical energy supply options, there is no motivation for trade; each region can be self-sufficient in carbon rights. Alternatively, if there are systematic differences among regions, the carbon tax will be determined by the least-cost region, and it will export carbon rights to the others. There is no need to construct an economic equilibrium model in order to analyze international trade in carbon rights during the later years. An intertemporal model is useful only during the transition period, when prices are first rising and then falling toward the backstop level. Moreover, an intertemporal model is needed in order to indicate the beginning date for the backstop era.

The no-trade carbon values may be compared with the time path required in order to equalize carbon prices in all regions. In general, the international equilibrium price is close to an average of the individual prices shown for the industrialized and the developing countries. The developing regions would not have large amounts of carbon available for sale until the backstop era beginning in 2050. Under the business-as-usual scenario, their domestic demands at that time would be just as large as those of the currently industrialized nations. During the backstop phase, the only major importer would be the Soviet Union. It is arbitrary which of the regions would supply carbon rights to the Soviet Union. For example, it could be assumed that their demands would be met by equal percentages from the carbon quotas of the other four regions.

Trade could lead to a large volume of international financial transfers. In 2020, for example, the United States would import carbon emission rights valued at about $50 billion. It would not be easy for political leaders to justify transfers of this magnitude. From the perspective of economic efficiency, however, this would be preferable to a detailed regulatory system in which specific carbon conservation guidelines are mandated for each class of end user.

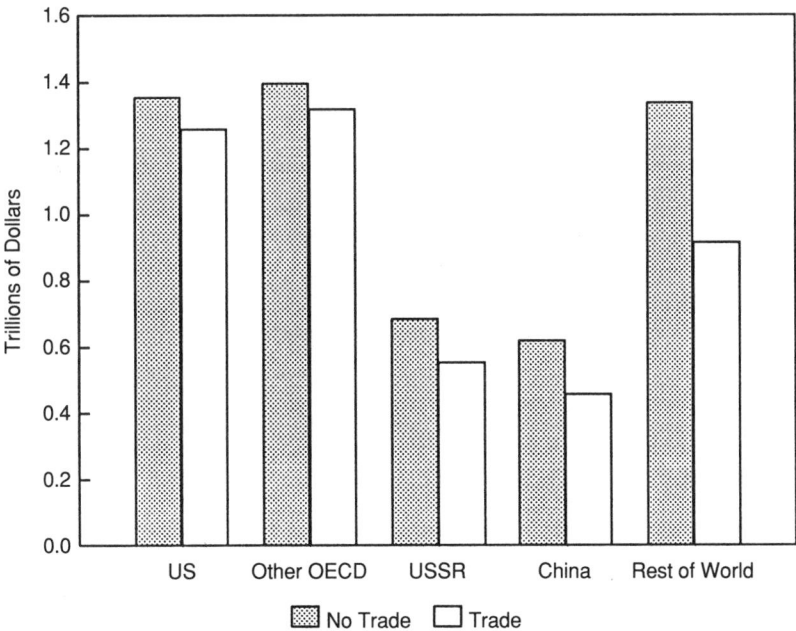

Figure 5.7 Consumption losses. Note: Discounted to 1990 at 5 percent per year.

What are the economic costs of carbon restrictions and what are the benefits from international trade? To smooth out year-to-year fluctuations, the results are summarized in terms of the impact upon discounted macroeconomic consumption throughout the planning horizon. Consumption losses are defined to be the difference between consumption with and without limits on carbon emissions. Employing a 5 percent real rate and discounting back to 1990, we obtain the results shown in figure 5.7. With no trade, it would cost the United States $1400 billion during the twenty-first century to participate in the international carbon limitation program. With trade, the costs would be lowered by about $100 billion.

From the viewpoint of economic efficiency, trade in emission rights is clearly worthwhile. We should not, however, expect dramatic cost reductions. Only small amounts of trade are needed to equalize the value of carbon emissions rights between regions. The import or export quantities rarely exceed 5 percent of the global emissions target.

In this case, trade does little to lower the overall cost of carbon constraints, but it would have an important indirect advantage. In the absence of trade in carbon emission rights, there could be distortions in the patterns of comparative advantage for energy-intensive basic materials. These would tend to be produced in countries where a low value is placed on carbon and would be exported to countries where a high penalty is imposed on emissions. Trade in carbon rights would enable the potential importing nations to avoid pressures from powerful domestic lobbies that would otherwise plead for protection against unfair foreign competition in energy-intensive materials such as metals and petrochemicals.

In these qualitative terms, it is straightforward to describe the indirect impacts of CO_2 limits on international trade in basic materials. Whalley and Wigle (1990) and Perroni and Rutherford (1991) have made a beginning in calculating the impact of carbon limits on trade, but much more remains to be done in this area.

Carbon Limits and the International Oil Market

Within Global 2100, the ROW region sets the international oil price and is the "swing" supplier of crude petroleum. (The ROW region includes not only OPEC but also Mexico and all other oil-exporting developing countries.) The United States and other OECD regions are viewed as price takers, and oil exports from the Soviet Union and China are subject to upper bounds. This is a convenient simplification but is not altogether satisfactory. An arbitrary price trajectory can lead to gaps in which the ROW region faces higher demands than it is prepared to supply.

There are interactions between carbon limits and the oil market. This is illustrated by figure 5.8, where the top line indicates the demands facing the ROW region under business as usual, and the bottom line refers to stabilization scenario I. With carbon taxes of about $200 per ton, there would be downward pressure on the international oil market. In effect, this would mean that a special-purpose tax of $30 per barrel is levied on crude oil by the consuming regions. Carbon limits would depress the demand for the ROW's oil exports.

Beyond 2030, nonelectric backstops could play an important role. Accordingly, we have eliminated supply-demand gaps in the international oil market by supposing that each region becomes self-sufficient from 2040 onward. The model suggests that carbon limits would impose severe restrictions on coal-based synthetic fuels as a bridge to the future. Conventional crude oil has an emissions coefficient only half

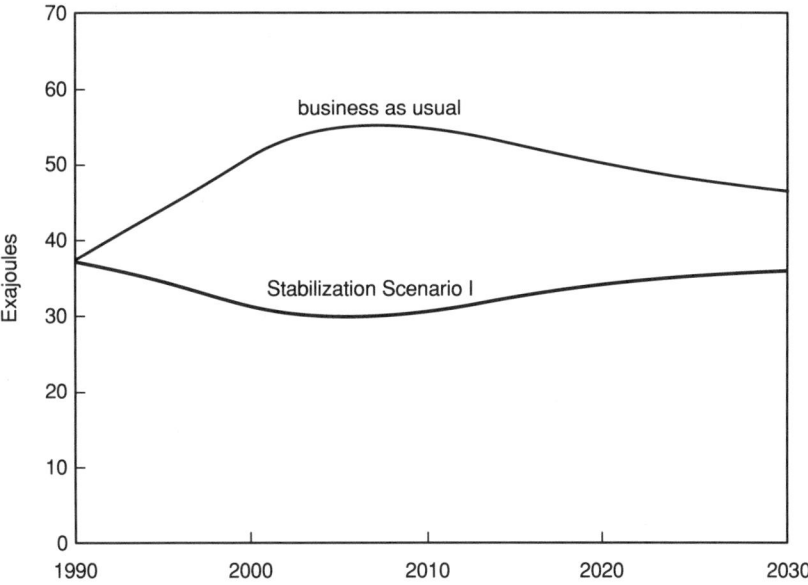

Figure 5.8   Rest of world oil exports

that of synthetic fuels and would therefore enjoy a premium value during the backstop phase.

In the absence of carbon limits, synthetic fuels impose a relatively low cap on the international oil price—perhaps $50 per barrel. With such limits, there would be an increased demand for oil relative to synthetics, and a higher international price is required in order to arrive at a market equilibrium. This means that the international value of petroleum could rise to $75 per barrel. CO_2 restrictions would impose a severe penalty on the use of coal-based fuels, and they could strengthen OPEC'S market power during the period when we might otherwise be shifting away from conventional oil.

An Alternative Emissions Stabilization Scenario

In order to arrive at an international allocation of emission rights, we would have to solve a number of equity-related issues. These are bound to create difficulties in an international negotiation process. In stabilization scenario I, we focused on economic efficiency. That is, for this specific set of CO_2 limits, we estimated their least-cost impact on individual regions. As a further step, it is useful to explore alternative

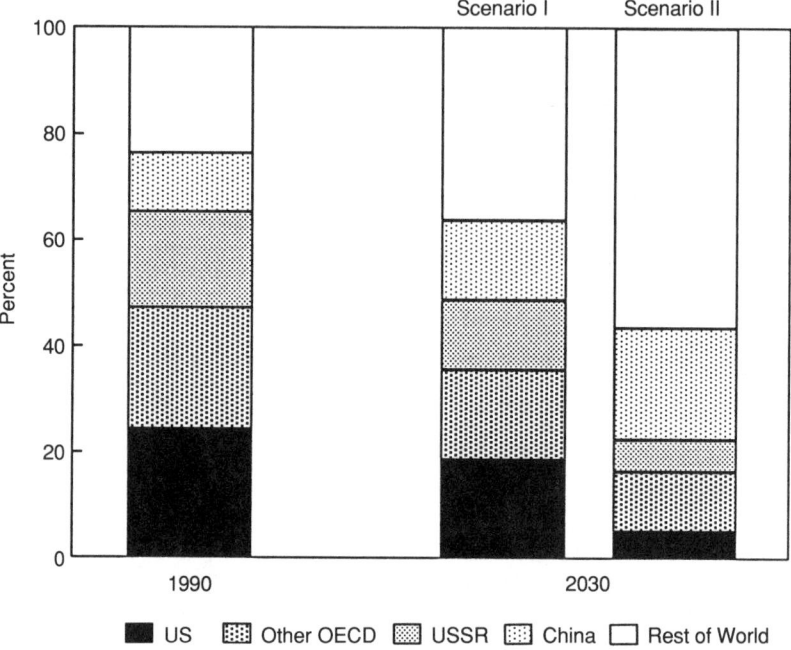

Figure 5.9 Carbon emission shares under two stabilization scenarios

allocation proposals and see how they might affect the sharing of international costs.

Stabilization scenario I is heavily influenced by the status quo. Scenario II leads to the identical level of global emissions but differs in how the burden might be shared among regions. Scenario II provides for a transition to a strictly egalitarian distribution of carbon rights. Initially, carbon rights are distributed among regions in proportion to their 1990 level of emissions. Over time, the shares change gradually. By 2030, carbon rights are distributed in proportion to 1990 population levels. The 1990 population base is chosen specifically so as to penalize nations that fail to control their rate of population growth.

Although both scenarios lead to a stabilization of global emissions, they differ significantly at the regional level. Figure 5.9 provides a comparison of carbon emission shares. Under stabilization scenario I, China and ROW are allotted 50 percent of global emissions in 2030, but their share rises to 80 percent under stabilization scenario II.

Such a major redistribution of emission rights would significantly alter the impacts on each region. Figure 5.10 shows the annual losses

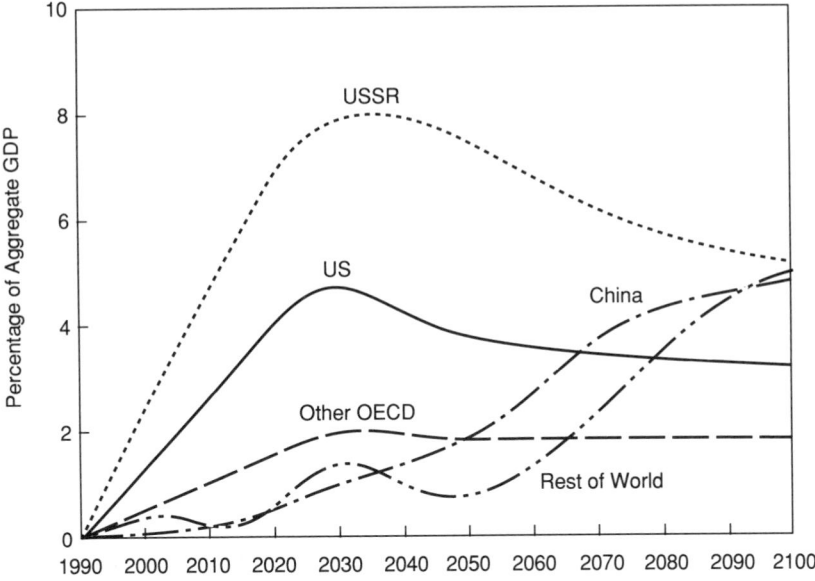

Figure 5.10 Annual losses under stabilization scenario II

as a percentage of GDP, and figure 5.11 compares the cumulative consumption losses, adding over all years and discounting to 1990 at 5 percent per year. There is no trade in carbon emission rights. Not surprisingly, the three industrialized regions are substantially worse off under scenario II. The reallocation of carbon rights leads to higher losses for them and lower losses for the developing countries. Scenario II may be more equitable, but the global losses are higher. This is not a constant-sum game.

The impact on the Soviet Union is particularly severe. It has huge natural gas resources, and under stabilization scenario I, these may be used as a low-cost alternative to coal-based synthetic fuels. Natural gas, however, is not carbon free. In order to comply with the targets required for stabilization scenario II, the Soviet Union would have to curtail its consumption of natural gas. The alternatives to natural gas are price-induced conservation and the nonelectric backstop, both of them costly. Moving from scenario I to scenario II will entail a significant GDP loss in the OECD regions but even higher losses in the Soviet Union.

More Rapid Growth in the Developing Countries

It is instructive to consider some of the surprises that might dramatically alter the costs of stabilizing emissions. In chapter 3, we provided sensitivity analyses for a single industrialized country, the United States. We examined how different perspectives on the costs and availability of supply and demand options affected cost projections. Technological optimism or pessimism would doubtless have similar impacts in other regions. One could identify combinations of assumptions about the potential for supply and demand side enhancements that would significantly alter the cost projections.

In the developing world, perhaps the biggest uncertainty relates to future GDP growth, which in turn depends on population and labor productivity growth. Our base case projections represent a considerable slowing over the rates experienced during the last few decades. It is a hazardous occupation to predict economic activity over a period spanning more than a century. Even looking one decade into the future, it is difficult to predict whether a particular region will experience rapid economic growth or stagnation. How sensitive are our results to these assumptions? What if the developing countries were to experience GDP growth rates one percentage point higher than those contained in our base case—certainly well within the range of current projections (Sathaye and Ketoff 1991).

Figure 5.12 compares the percentage GDP losses for the base case and the accelerated GDP scenario. We continue to assume that the goal is stabilization of global emissions. In this example, we revert to stabilization scenario I for defining the international allocation of carbon rights.

Losses are quite sensitive to GDP assumptions. The developing countries would gain from more rapid growth, but their percentage GDP losses would rise more rapidly. They would then become purchasers rather than sellers of carbon emission rights. In an unconstrained scenario, they would have even higher emissions than in the base case. If the developing countries enjoy high growth, including them in an international agreement would be even more important—and also more costly.

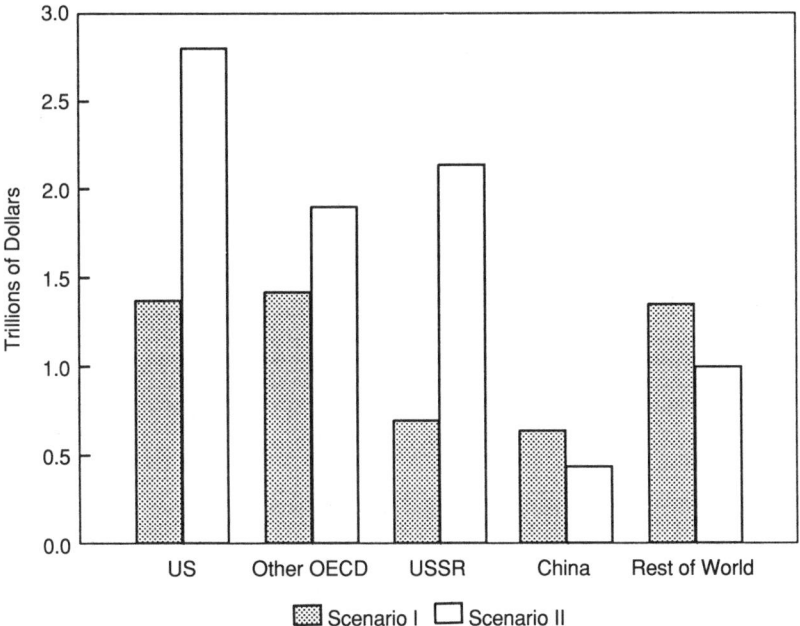

Figure 5.11 Consumption losses under alternative stabilization scenarios. Note: Discounted to 1990 at 5 percent per year.

A More Drastic Carbon Limit

Up to this point, we have focused on stabilizing global emissions at 1990 levels, but stabilizing emissions is not the same as stabilizing atmospheric CO_2 concentrations. Houghton et al. (1990) report that stabilizing emissions at current levels will only slow the rate at which atmospheric CO_2 concentrations increase. By the end of the twenty-first century, concentrations would still be twice their preindustrial levels.

Suppose that after further research on the possible impacts of climate change, it is agreed that a doubling in atmospheric CO_2 concentrations will entail significant economic, environmental, and social consequences. A doubling might be delayed (or even prevented) through more aggressive measures to control emissions. For example, Houghton et al. (1990) report that CO_2 concentrations could be stabilized at 1990 levels by an immediate cut of 70 percent from current levels. Further cuts would be required by 2050.

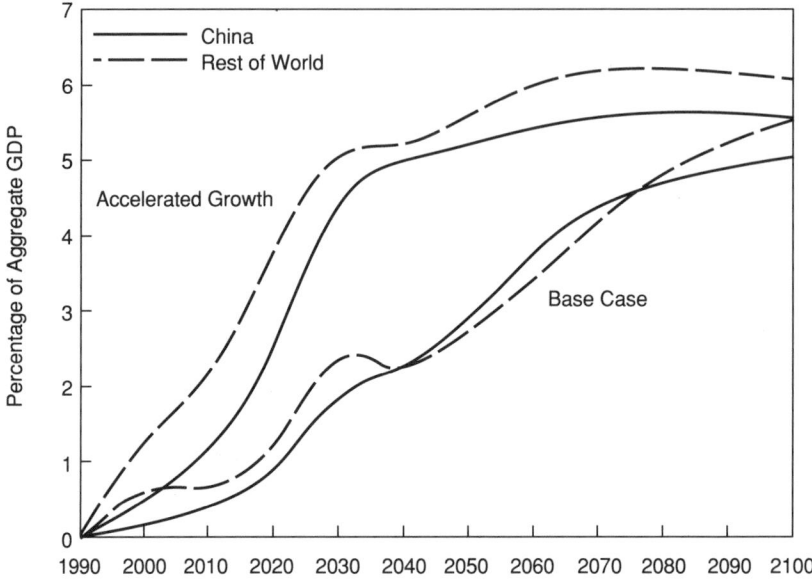

Figure 5.12   Annual losses in developing countries under two growth scenarios

An economically rational public policy would aim for maximum net benefits, including both public and private costs. If drastic emissions cuts are contemplated, it is important to understand not only what these cuts might buy in terms of reducing the rate of climate change but also how much they might cost. Our next scenario is designed to provide some insights into how the control costs might rise with more stringent emission limits than those considered in stabilization scenarios I and II.

Consider a case in which emissions are to be reduced 50 percent below 1990 levels, with the goal to be achieved by 2030. Although this would not stabilize atmospheric CO_2 concentrations, it would delay the date at which we reach the doubling point above preindustrial levels. For burden sharing purposes, we assume that initially (during the year 2000), carbon rights are distributed among regions in the same proportions as in 1990, but that by 2030 carbon rights are distributed among regions in proportion to their 1990 level of population.

Figure 5.13 shows the discounted consumption losses that would accrue under this 50 percent emissions cutback and compares them with those under stabilization scenarios I and II. The tighter constraints lead

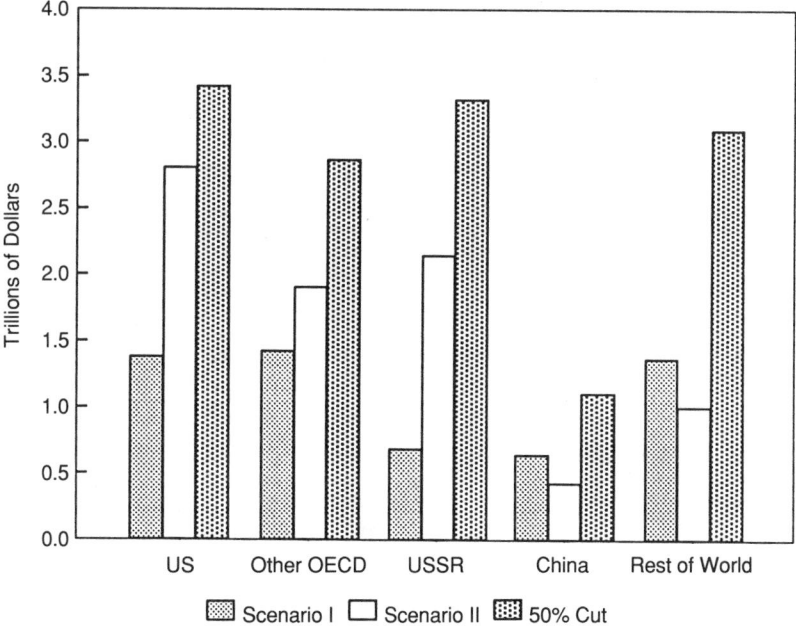

Figure 5.13 Consumption losses under alternative emissions scenarios. Note: Discounted to 1990.

to substantially higher losses in each region. The magnitude of the increase is quite sensitive to the point of comparison. For industrialized countries, the cost will be much higher when the 50 percent cutback is compared to stabilization scenario I. For developing countries, the reverse is true.

Figure 5.14 compares carbon allotments in 2030 under all three scenarios. For the industrialized countries, stabilization scenario II accomplishes most of the reduction required under the 50 percent global emissions target. The additional reduction is relatively small. Conversely, developing countries have higher allotments under stabilization scenario II. They would incur greater additional costs under the 50 percent global emissions cut.

Summary and Conclusions

This chapter has provided a series of sensitivity analyses. We have tried to show how the costs of carbon emissions limits might differ among regions. Although there is enormous uncertainty inherent

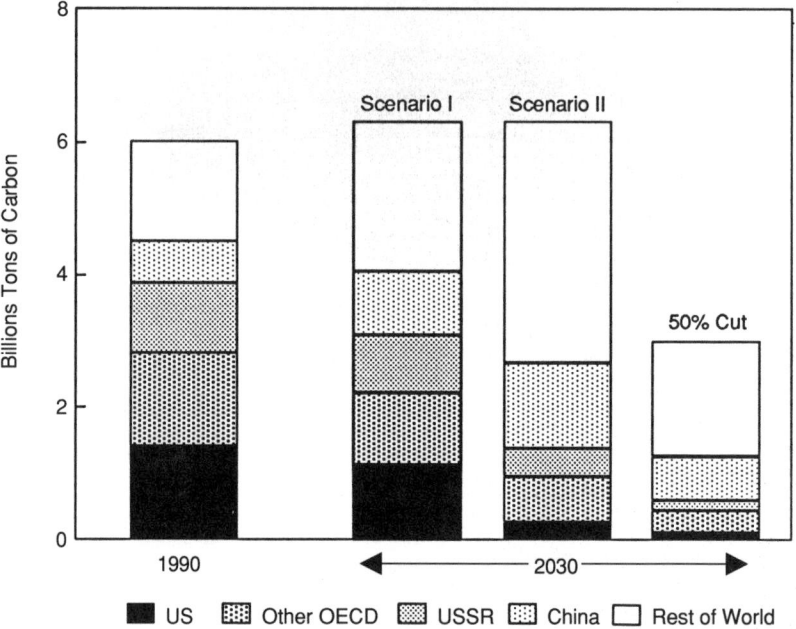

Figure 5.14   Carbon allotments under alternative allocation schemes

in projections of this type, nevertheless, we believe that the Global 2100 calculations suggest some general results. In the absence of an international agreement, carbon emissions are likely to increase considerably—perhaps by a factor of seven over the next century. During this period, there is likely to be a significant shift in the regional pattern of emissions. In 1990, the industrialized countries accounted for 64 percent of the total. By 2100, their share could drop to 30 percent or less.

When considering global emission reductions, international negotiators will be confronted with a number of formidable issues. It is all too easy for diplomats to reach agreement on no-regrets policies, but eventually they will have to confront the difficult questions. By how much should global emissions be reduced? How should the reductions be allocated? Who should pay?

Ideally, global emission goals should be determined so as to balance incremental benefits and incremental costs. To date, little effort has been made to arrive at such a balance. We continue to debate over arbitrarily chosen emission targets. Given the uncertainty surrounding the economic, environmental, and social consequences of global warming, it is no surprise that the international community has failed to arrive at a consensus.

This book does not attempt to determine the optimal amount of emission reductions. Rather, we examine the costs of alternative CO_2 scenarios. Under our base case, using assumptions that we believe are reasonable, the Global 2100 analysis indicates that stabilizing emissions at or near current levels could be quite costly.

The costs are likely to be quite sensitive to a variety of factors. The loss estimates will depend on the cost and availability of carbon-free supply and demand options. The losses will also vary with the GDP growth assumptions, the regional allocation of emission rights, and the existence or the absence of an international market in emission rights.

Costs are sensitive to the absolute level of the carbon emissions constraint. In the next chapter, we will see that a halving of the emissions limit does not necessarily lead to a doubling of the overall costs. This depends on the shape of the willingness-to-pay functions.

References

M. Grubb. 1989. *The Greenhouse Effect: Negotiating Targets*. The Royal Institute of International Affairs.

J. T. Houghton, G. J. Jenkins and J. J. Ephraums. 1990. *Climate Change—The IPCC Scientific Assessment*. Cambridge University Press, Cambridge.

C. D. Masters, E. D. Attanasi, W. D. Dietzman, R. F. Meyer, R. W. Mitchell, and D. H. Root. 1987. "World Resources of Crude Oil, Natural Gas, Natural Bitumen, and Shale Oil." Twelfth World Petroleum Congress, *Proceedings*, vol. 5, pp. 3–27.

National Academy of Sciences. 1991. *Policy Implications of Greenhouse Warming—Synthesis Panel*. National Academy Press, Washington, D.C.

C. Perroni and T. Rutherford. 1991. "International Trade in Carbon Emission Rights and Basic Materials: General Equilibrium Calculations for 2020." Wilfrid Laurier University and University of Western Ontario, April.

J. Sathaye and A. Ketoff. 1991. "CO_2 Emissions from Major Developing Countries." *Energy Journal* 12, (1), January.

J. Whalley and R. Wigle. 1990. "The International Incidence of Carbon Taxes." University of Western Ontario and Wilfrid Laurier University, October.

6

Willingness to Pay for Carbon Emission Rights

Small-Scale Examples

Much of this book can be summarized in terms of willingness-to-pay functions—the marginal value of carbon associated with each possible level of emissions curtailment. To understand some of the results produced by Global 2100, it is useful to turn to simpler models, analyzing a single point in time and ignoring the dynamics of the transition away from exhaustible hydrocarbon resources.

In constructing figure 6.1, we begin with a scenario in which no constraints are imposed on carbon emissions. The horizontal axis measures x, the level of emissions as a fraction of those in the business-as-usual scenario. The vertical axis measures the willingness to pay for carbon rights—and the associated losses in conventionally measured GDP. Following the shape suggested by figure 5 in Nordhaus (1991), the GDP losses (dashed line) are shown as a quadratic function. When no controls are imposed, (x = 100 percent), the losses are zero. At this point, the marginal value of emission rights is zero, and the function has a slope of zero. When cutbacks are required, the loss function summarizes the ability of the economy to undertake an efficient mix between price-induced conservation and supply-side shifts in the composition of fuels. The greater is the reduction, the more will one be willing to pay to avoid further cutbacks.

It is convenient to measure GDP in units so that a zero level of emissions leads to one unit of losses. With a quadratic GDP loss function, the marginal losses are a linear function of x. At zero emissions, the marginal costs are twice the average costs. The marginal cost function may also be interpreted in terms of willingness to pay. It indicates the level of carbon taxes that would have to be imposed in order to reach a given target for curtailment.

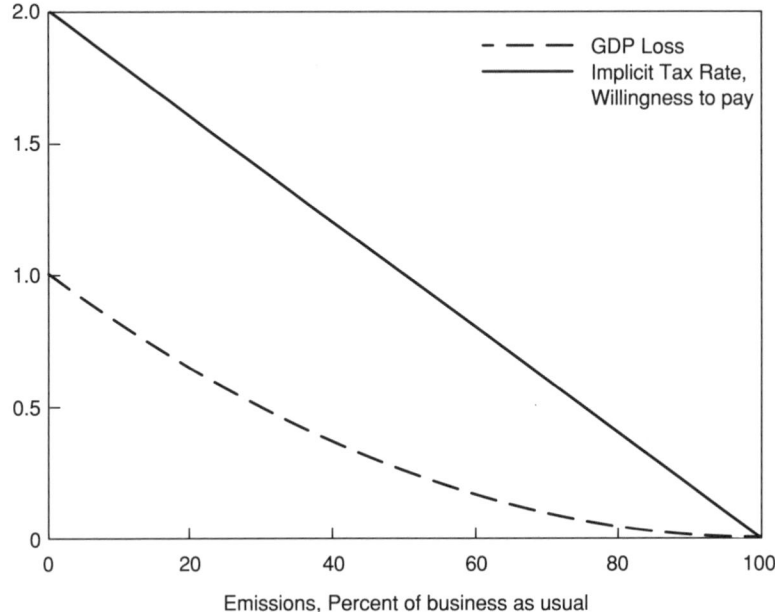

Figure 6.1 Linear willingness to pay: quadratic GDP loss function

Business as usual could eventually lead to a sevenfold increase over today's CO_2 level (recall figure 5.1). Suppose the carbon limit were set at 20 percent of business as usual during the late twenty-first century. This would mean that emissions were stabilized at roughly their current levels. Now consider two alternative scenarios. In one, the limit is placed at 20 percent of business as usual; in the other at 10 percent.

From figure 6.1, we can see that small levels of curtailment would require low tax rates and would entail almost negligible GDP losses. The figure also helps us to understand two types of numerical results from the Global 2100 model: First, the losses are higher when emissions are limited to 10 percent rather than 20 percent, but they do not double. With a quadratic function, the losses depend on the level of curtailment relative to the future business-as-usual scenario, not relative to the current level. Second, for similar reasons, the carbon tax rate (the marginal cost of reduction) does not double when we move from the 20 percent to the 10 percent scenario. For short, we may refer to those two generalizations as insensitivity results.

Insensitivity follows directly from the quadratic loss function and the observation that proposed CO_2 limits are often less than half the

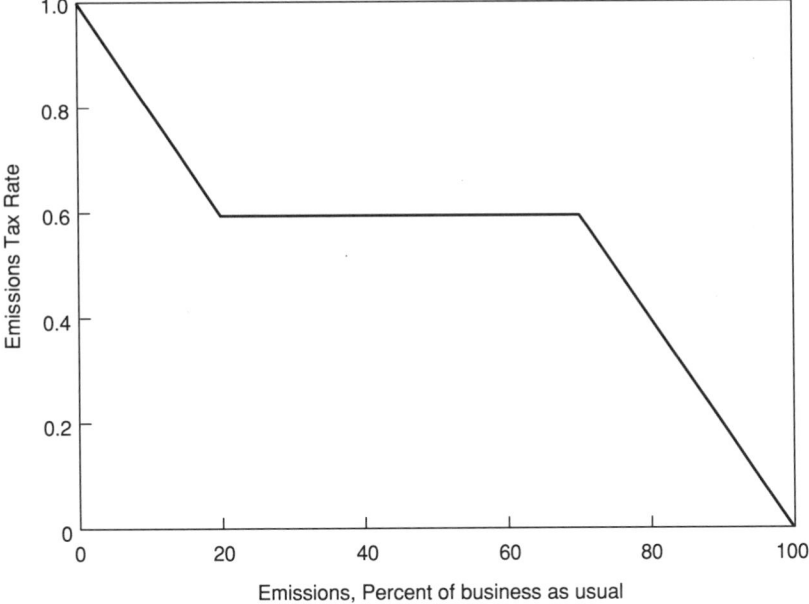

Figure 6.2 Nonlinear willingness to pay: effect of backstop technologies

business-as-usual level in the distant future. But what if the loss function does not follow a quadratic form? Suppose, for example, that at low emission levels, the losses are hyperbolic in form—proportional to $1/x$. In this case, both the average and the marginal costs of control would increase without limit as x approaches zero. With a hyperbolic loss function, we could not expect insensitivity.

As it turns out, Global 2100 displays even less sensitivity than might be anticipated from a quadratic GDP loss function. Consider the nonlinear willingness-to-pay function shown in figure 6.2. This coincides with figure 6.1 at emission levels between 70 percent and 100 percent of business as usual. Within this range, a high level of emissions is associated with a low willingness to pay and a low marginal value of carbon. At high emission levels, it is straightforward to observe that there is a low cost for low levels of curtailment.

Once emissions are curtailed to 70 percent of business as usual, there is a plateau in the willingness-to-pay function at 0.6. This is the tax rate at which consumers would face identical costs for the carbon-based and the carbon-free backstops. On figure 6.2, the two technologies would both be utilized if the permissible limit lies anywhere between

20 percent and 70 percent. In order to curtail emissions below 20 percent of business as usual, the tax would have to be raised above the plateau level. The carbon-based backstop would then no longer be competitive with the carbon-free source of supply, and further reductions would become increasingly expensive.

Figures 6.1 and 6.2 provide some insights on the potential gains from international trade. Suppose there are only two regions in the world, called North and South. They have identical willingness-to-pay functions. The only difference is that North's CO_2 quota is 30 percent of its business-as-usual level and South's is 70 percent. With the quadratic loss function of figure 6.1, it can be shown that the equilibrium quantity traded is 20 percent of each region's business-as-usual level. Since the equilibrium carbon price happens to be 1.0 in this example, the value of imports and exports is .20 units' worth of GDP. Both parties would benefit from trade, and their gains may be determined as follows:

Region	Losses (no trade)	Losses (with trade), excluding value of carbon imports or exports	Losses (with trade)	Regional gains from trade
North	.49	.25	.25 + .20 = .45	.04
South	.09	.25	.25 − .20 = .05	.05

With identical quadratic loss functions and different quota levels, there are substantial gains from trade. Now continue with the assumption that North and South have the same GDP loss functions but that this function is obtained by integrating the area beneath the nonlinear willingness-to-pay function shown in figure 6.2. If the North and South carbon quotas remain at 30 percent and 70 percent, respectively, each region would be in equilibrium at the backstop tax rate of .6. With identical backstop technologies, there would be identical carbon prices in both regions and no gains from trade.

The plateau in figure 6.2 provides a partial explanation of why there are such low gains in the Global 2100 model. In addition, it is worth recalling that the five regions are quite unequal in size. Through 2020, the business-as-usual carbon demands would continue to be considerably larger in the industrialized regions than in China and the ROW. Stabilization scenario I does not provide the developing countries with sufficiently large quotas to satisfy the industrialized nations' demands for carbon at low prices.

Results from Global 2100

The small-scale examples are instructive, but they do not explain the dynamics of the CO_2-energy-economy system. In Global 2100, the willingness-to-pay functions are heavily influenced by the time lags in demand responses and by the time required for new supply technologies to penetrate into the market. To see how the model might deal with alternative reduction scenarios, we have run a series of numerical experiments in which all regions maintain a uniform carbon tax throughout the twenty-first century. The rate of this tax is varied from zero up to $1000 per ton but is not changed from one time period to the next. All revenues are redistributed internally within each region.

Figure 6.3 summarizes the results for 2020 (dashed line) and 2100 (solid line). The two dates mark the beginning and the end of the transition period away from conventional hydrocarbons. The figure demonstrates the importance of short-run rigidities and long-run substitution. In 2020, it is too early for the backstops to have any influence on the ability of the world economy to respond to alternative levels of emissions control. There are no plateaus, and the results look much like the linear willingness-to-pay function illustrated in figure 6.2.

By contrast, the backstops would play a major role in 2100. The coal-based synthetic fuels technology accounts for virtually all of the carbon released in the business-as-usual scenario, for there would no longer be any limits on its rate of market penetration. With a small but positive carbon tax, there would be only moderate reductions in carbon emissions. No discontinuities occur until the tax reaches the plateau of $208 per ton. At that point, according to our base technology assumptions, the nonelectric carbon-free backstop becomes competitive, and coal-based synthetic fuels could be eliminated altogether. This would enable a drastic reduction in global carbon emissions. Further reductions are feasible but would become increasingly expensive.

Concluding Comments

There are many limitations inherent in this numerical experiment. In particular, there is no reason to believe that it would be economically

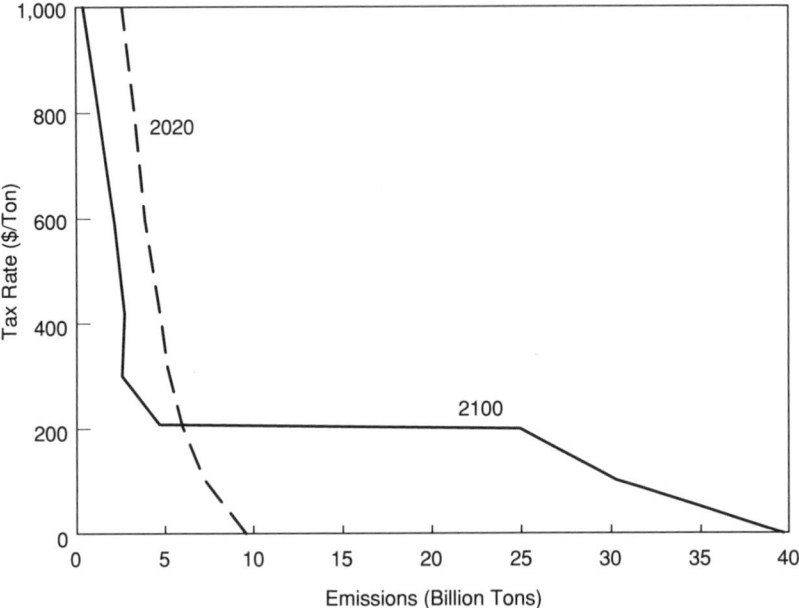

Figure 6.3 Willingness to pay, 2020 and 2100

efficient to maintain a uniform carbon tax throughout the twenty-first century. Despite the difficulties, there is a clear message from the analysis. If there are price plateaus associated with backstop technologies, policy makers must be aware of this possibility. Given sufficient time for the market penetration of new technologies, it could happen that small changes in tax rates would lead to large changes in emission levels.

Should threshold phenomena be important, there is a strong argument for implementing national quotas by auctioning off carbon rights rather than attempting to set tax rates. Emission levels would be far more predictable if governments were to operate directly on quantities, and they allowed the market process to determine the price at which these quantities are to be sold.

Reference

W. Nordhaus. 1991. "The Cost of Slowing Climate Change: A Survey," *Energy Journal* 12, (1), January.

II Model Structure

7 Global 2100: Model Formulation

Intertemporal Model Structure

This chapter describes the basic structure of Global 2100, a model of CO_2-energy-economy interactions. In its simplest form, the analysis is based on parallel independent computations for five major geopolitical regions: the United States, other OECD nations, the Soviet Union, China, and ROW (the rest of the world). Each region must live within an independently specified carbon emissions quota. In chapter 8, we show how the model can be extended to the case of international trade in carbon emission rights.

Global 2100 is benchmarked against a base year of 1990, and the projections cover ten-year time intervals extending from 2000 through 2100. The model is intertemporal rather than recursive. It is assumed that producers and consumers will be sufficiently forward looking to anticipate the scarcities of energy and the environmental restrictions that are likely to develop during the coming decades. For an example of the recursive approach, see Edmonds and Reilly (1985).

For each region, a dynamic nonlinear optimization is employed to simulate either a market or a planned economy. There is a single representative producer-consumer. Supplies and demands are equilibrated within each individual time period, but there are "look-ahead" features to allow for interactions between periods. Since "putty-clay" assumptions are built into the model, price responsiveness is lower in the short run than over the long run. That is, initial capital stocks are viewed as hard-baked "clay," and subsequent investments are malleable "putty."

During each period, it is supposed that energy consumers are locked in by their past acquisitions of capital goods and by the rate of depreciation. There is flexibility only in the case of new investment. This is where the model provides for substitutability between energy, capital, labor, and output. For new vintages of investment, the energy-output ratios are determined not only by current prices but also by the anticipation of future price changes.

A fully intertemporal model seems like an unrealistic description of the way that economic agents react to market signals. Why not assume static expectations—that future prices will remain the same as current prices? Why bother with the implausible assumption of perfect foresight? An answer was provided by Solow (1974) in his Ely Lecture to the American Economic Association: "If a market-guided system is to perform well over the long haul, it must be more than myopic. Someone—it could be the Department of the Interior, or the mining companies, or their major customers, or speculators—must always be taking the long view. They must somehow notice in advance that the resource economy is moving along a path that is bound to end in disequilibrium of some extreme kind."

The foresight assumption affects both demands and supplies in Global 2100. Much as in the Hotelling literature, there are expectations that influence the time path and the extraction profile of exhaustible resources. (Exhaustible resources are defined here to include not only conventional hydrocarbons but also the rights to emit CO_2.) Expectations affect the willingness to incur immediate costs for speeding up the rate of market penetration of new supply technologies. This in turn affects the extent of overshoot beyond backstop levels of energy costs.

Expectations affect the accumulation of capital over time. Savings and investment decisions are modeled through the maximization of discounted utility. Just as in the Ramsey model of optimal economic growth, consumers will receive equal benefits from an additional dollar's worth of current consumption and the future consumption generated by an additional dollar's worth of investment.

At some point, it would be instructive to take the results from an intertemporal model such as Global 2100 and compare them with a dynamic model that is similar except that it has a recursive structure—that is, is based on the assumption of static expectations. On the energy demand side, one such comparison is reported in chapter 9.

Formulation of a Single-Region Model:
Sets, Variables, and Data Tables

To decompose the overall problem into more manageable subproblems, it is supposed that each of the five regions faces an exogenously determined carbon emissions quota. In the case of international trade in crude oil, the model is almost consistent. It is assumed that the ROW (which includes OPEC) sets an international price, the OECD nations are price takers, and the ROW region meets their demands for net imports. Alternative scenarios (informal iterative methods) are employed to eliminate the prospective gaps between oil supplies and demands. We have not attempted to define a procedure for changing OPEC prices in response to changes in the value of carbon emission rights.

Within each region, the analysis is based on ETA-MACRO, a model of two-way linkage between two submodels: ETA, an energy technology assessment model, and MACRO, a macroeconomic production function describing the balance of the economy. (Recall figure 2.2.) Energy supplies include both exhaustible hydrocarbon resources and also backstop technologies. Associated with each of the supply technologies are coefficients describing the costs and the carbon emissions per unit of the activity level. There are upper bounds on the speed of introduction of new technologies and lower bounds on the rates of decline of obsolete ones.

In order to avoid the data-intensive approach associated with an end-use analysis, energy demands are divided into just two categories: electric and nonelectric. Along with capital and labor, energy is viewed as a basic input into the economy-wide macro production function. The growth rate of the labor force (measured in "efficiency units") determines the potential rate of GDP growth within each region. These rates constitute one of the key inputs into the model.

Energy consumption and carbon emissions need not grow at the same rate as the GDP. Over the long run, they may be decoupled from GDP growth. In ETA-MACRO, these possibilities are summarized through three parameters: AEEI (autonomous energy efficiency improvement), ESUB (the elasticity of price-induced substitution between energy and other productive inputs), and PNREF (the reference price employed in benchmarking each region's production function). During a period of rising international energy prices, both price-induced and nonprice conservation will permit a significant reduction in energy demands. Under most scenarios, however, they are not sufficient to stabilize total primary energy consumption. Additional new sources of energy supplies are needed in both the electric and nonelectric sectors.

Global 2100 is written in GAMS, a generalized algebraic modeling system (Brooke et al. 1988). Solving two cases (with and without a carbon constraint) for all five regions in parallel takes 15 minutes on a 25 MHz desktop 486 computer equipped with a coprocessor. With the GAMS language, it is natural to begin by describing the sets, then the decision variables, and the data tables. A typical data set is provided in appendixes A through E. The sets are as follows:

ET   Electric energy supply technologies: /HYDRO, GAS-R, OIL-R, COAL-R, NUC-R, GAS-N, COAL-N, ADV-HC, ADV-LC/

NT   Nonelectric energy supply technologies: /OIL-MX, CLDU, OIL-LC, GAS-LC, OIL-HC, GAS-HC, SYNF, RNEW, NE-BAK/

X   Exhaustible hydrocarbon resources: /OIL-LC, GAS-LC, OIL-HC, GAS-HC/

DL   Electricity technologies subject to decline limits: /GAS-N, COAL-N, ADV-HC/

EX   Electricity technologies subject to expansion limits: /ADV-HC, ADV-LC/

NX   Nonelectric technologies subject to expansion limits: /SYNF, RNEW, NE-BAK/

GS   Natural gas supply and demand technologies: /GAS-R, GAS-N, GAS-LC, GAS-HC/

Among the decision variables, the maximand UTILITY is a scalar. All other variables are time indexed and refer to the projection periods $t = 1(2000), 2(2010), \ldots, T(2100)$. Base year values are denoted by $t = 0(1990)$. Upper and lower bounds on all individual variables are shown in box 7.1. For simplicity, the time index t is omitted from the variables listed below:

UTILITY   Sum of discounted logarithms of aggregate consumption

Units of measurement for the following variables are $ trillion per year (measured in dollars of constant 1990 purchasing power):

C   Consumption
I   Investment
EC   Energy costs
Y   Production, excluding energy sectors
YN   New vintage production, excluding energy sectors

Box 7.1
Bounds on Individual Variables

Lower bounds are imposed on almost all of the variables. Some of the lower bounds are zero. Others help to reduce the solution time, and still others prevent program calls for undefined numbers, (for example, for the logarithm of zero). Some of these lower bounds are essential during intermediate iterations but are intended to be nonbinding constraints at an optimal solution.

It may happen that units of measurement are chosen so that the logarithm of consumption is negative. In other cases, energy costs may be negative. That is, oil export revenues exceed domestic energy costs. To allow for these unusual possibilities, no lower bounds are assigned to the variables UTILITY and EC. They are allowed to take on negative as well as positive and zero values.

Upper bounds are imposed on the variables PE and PN through the data tables ECAP and NCAP, respectively. There are additional upper and lower bounds on individual PE and PN variables:

Production is fixed at the values shown in the ECAP table for the following types of electrical generating facilities: HYDRO and NUC-R.

The CLDU (coal—direct use) variables are subject to an upper bound determined by their base year level, the potential GDP growth rates, and the CLGDP elasticity specified in the MACRO table.

The OIL-MX (oil import minus export) variables are subject to the lower and upper bounds specified in the tables OILM and OILX. The ROW region includes OPEC. For this region, there is an additional bound on exports: the net demands generated by the United States and other OECD countries less the amounts supplied by the Soviet Union and China. This means that the ROW solution must be calculated after those for the other four regions.

The CARM and CARX (import and export variables for carbon rights) are nonnegative and subject to the upper bounds specified in the tables CARML and CARXL. When there is no international trade in such rights, these upper bounds are set at zero.

Units of measurement for the following variables are $ trillion:

K   Capital stock

KN   New vintage capital

Units of measurement for the following variables are TkWh (trillion kilowatt-hours) per year:

E   Demand for electric energy before AEEI

EN   New vintage demand for electric energy

PE   Supply of electric energy, members of the set ET

XPE   Above-normal expansion of electric energy, members of the set EX

Units of measurement for the following variables are exajoules (10^{18} joules) per year:

N   Demand for nonelectric energy before AEEI

NN   New vintage demand for nonelectric energy

PN   Production of nonelectric energy, members of the set NT

XPN   Above-normal expansion of nonelectric energy, members of the set NX

GN   Gas consumed to meet nonelectric energy demands

RA   Gross additions to proved reserves

Units of measurement for the following variables are exajoules (10^{18} joules) :

RSC   Undiscovered resources

RSV   Proved reserves

Units of measurement for the following variables are billion metric tons of carbon per year:

DC   Quantity of carbon deferred from use in period t and available in period t + 1.

CARM   Quantity of carbon imported from other regions

CARX   Quantity of carbon exported to other regions

In formulating the constraints, all parameters have been indicated by lowercase letters. These values are specified either directly or indirectly through a series of data tables (which users are free to modify). In appendix A, for example, the MACRO table contains the region-by-region values for gdp_0 (the initial GDP), kgdp (the initial capital-GDP ratio), kpvs (capital's value share), and depr (the annual depreciation rate). The potential GDP growth rates ($grow_t$) are contained in the GROW table. All input data are grouped into the following files:

MACRO	MACRO	Macroeconomic and other parameters
	GROW	Potential GDP growth rates
	AEEI	Autonomous energy efficiency improvements
ELEC	ECAP	Electricity production capacities
	ECST	Electricity cost coefficients
NELE	NCAP	Nonelectric production capacities
	NCST	Nonelectric cost coefficients
	SDAT	Supply data—exhaustible hydrocarbon resources
OILTRADE	OILP	International oil price scenarios
	OILX	Oil export limits
	OILM	Oil import limits
CARBON	CARP	International carbon prices
	CARLIM	Annual carbon limits
	CH	Carbon emission coefficients and heat rates
	CARXL	Carbon export limits
	CARML	Carbon import limits

MACRO Constraints

Within each region, there is a single equation to define the maximand UTILITY and a single constraint referring to the terminal period, TC. All other constraints are time indexed and refer to the projection periods $t = 1(2000), 2(2010), \ldots, T\,(2100)$. They may be grouped as follows:

Macro constraints

UTIL   Discounted utility, sum over all projection periods

CC	Allocation of total output capacity among expenditure categories
NEWCAP	New capital stock, excluding energy sector
TC	Terminal condition on investment and capital stock

Putty-clay constraints

NEWPROD	New vintage production, excluding energy sector
TOTALPROD	Total production, excluding energy sector
TOTALCAP	Total capital stock, excluding energy sector
NEWELEC	New vintage demand for electric energy
NEWNON	New vintage demand for nonelectric energy

ETA constraints

SUPELEC	Supply of electric energy
SUPNON	Supply of nonelectric energy
SUPGAS	Supply of gas

EXP(EX)	Maximum expansion rate of EX (electric) technologies
NXP(NX)	Maximum expansion rate of NX (nonelectric) technologies
DEC(DL)	Maximum decline rate of DL (electric) technologies

PRVLIM(X)	Production-reserve ratio, exhaustible hydrocarbon resources
RSVAV(X)	Proved reserves available
RSCAV(X)	Undiscovered resources available
RDFLIM(X)	Resource depletion limit
ANC	Annual carbon limit

COSTNRG	Cost of energy

The macro constraints begin with the UTILITY maximand:

$$\text{UTIL: UTILITY} = \sum_{t=1}^{T-1}(\text{udf}_t)(\log C_t)(10) + (\text{udf}_T)(\log C_T)[5 + (\text{udr}_T)^{-1}],$$

where the utility discount rate for period $t = \text{udr}_t = (\text{kpvs/kgdp}) - \text{depr} - \text{grow}_t$, and the utility discount factor for period $t = \text{udf}_t = \prod_{\tau=0}^{t-1}(1 - \text{udr}_\tau)^{10}$. The factors of 10 allow for the fact that the first $T-1$ periods are each 10 years in length. The terminal period begins five years prior to the terminal date and extends an infinite length of time thereafter. This is the reason for the factor of $[5 + (\text{udr}_T)^{-1}]$.

A numerical example shows how the utility discount rate is determined if the following parameter values are adopted:

kpvs = capital's value share   = 24%
kgdp = initial capital/GDP ratio = 2.4 years
depr = depreciation rate   = 5%/year

net rate of return on capital   = (24%/2.4 years) − 5%/year
                                     = 5%/year
$grow_t$ = potential growth rate   = 2%/year
∴ udr_t = utility discount rate   = 3%/year

The utility discount rates are chosen for descriptive rather than normative purposes. With the logarithmic single-period utility function, these values ensure that the optimal steady-state growth rate will coincide with that assumed for the potential GDP. Along an optimal path, the rate of decline in the present value of the marginal utility of consumption will equal the net marginal productivity of capital. (For a calculus-of-variations proof of this proposition, see Chakravarty 1969, p. 65.) Moreover, these discount rates mean that the economy-wide savings rate will adjust downward (upward) automatically if there is a drop (rise) in the potential GDP growth rate.

The CC equations specify that the gross value of production is to be divided among current consumption, investment for building up the stock of capital, and interindustry payments for energy costs:

$$CC_t: \quad Y_t = C_t + I_t + EC_t \qquad t = 1, \ldots, T$$

Since the variable C_t enters only into the objective function and into constraint CC_t, the dual variable for constraint CC_t may be interpreted as the present value of the marginal utility of consumption during period t. First-order optimality conditions lead to the Ramsey rule for the optimal allocation over time among savings, investment, and consumption. That is, the marginal productivity of capital determines the rate of decline of these dual variables from one period to the next. All other dual variables for period t have a similar interpretation; they are present value prices. In order to convert them into future values, they must be divided by the dual variables for the CC_t constraints. According to the numerical example, the net marginal productivity of capital is 5 percent, and the dual variables for the CC_t constraints would decline by about 5 percent annually.

The NEWCAP equations describe the dynamics of capital accumulation. Within each 10-year period, net new capital formation is determined by gross investment less depreciation. Since investment is measured as an annual flow, an accumulation factor of 5 is applied to the beginning and ending rate of investment to determine net new capital formation during the decade as a whole:

$$\text{NEWCAP}_{t+1}: \quad \text{KN}_{t+1} = 5[(1 - \text{depr})^{10} \, I_t + I_{t+1}] \qquad t = 0, \ldots, T - 1,$$

where $I_0 = (\text{grow}_0 + \text{depr})(\text{kgdp})(\text{gdp}_0)$.

At the end of the planning horizon, a terminal constraint is applied to ensure that the rate of investment is adequate to provide for replacement and net growth of the capital stock during the subsequent periods. This reduces horizon effects, but is not guaranteed to eliminate them entirely. For a more complete discussion of terminal conditions, see Svoronos (1985):

$$\text{TC}_T: \quad \text{K}_T(\text{grow}_T + \text{depr}) \leq I_T$$

To allow for time lags in the response of energy demands to changing prices, there is a putty-clay model of aggregate production. Because energy consumption is closely associated with long-lived capital goods (such as residential housing), a sharp distinction is drawn between ex ante and ex post substitutability. That is, there is full substitutability between productive inputs at the time that new vintages of capital stock are generated. The optimal choice of input mix is determined not only by current but also prospective prices. Once the mix is determined, no additional opportunities are provided for revising the input-output coefficients for vintage t. The only flexibility is associated with new investment for replacement and expansion of the capital stock during later periods.

The output of vintage t is determined by a nested CES (constant elasticity of substitution) production function. (For a comparison of alternative forms of nesting, see box 7.2.) The first term in the nest indicates that capital and labor may be substituted for each other—for example, through automation of labor-intensive tasks. The higher is the wage rate, the more attractive it becomes to adopt automation. Similarly, the second term indicates that electric and nonelectric energy may be substituted for each other—for example, electric-driven heat pumps in place of oil- or gas-fired furnaces. The higher is the price of nonelectric energy, the more attractive it becomes to adopt heat pumps. The new vintage production function is of the following specific form:

$$\text{NEWPROD}_t: \quad YN_t = [a(KN_t)^{\rho\alpha}(LN_t)^{\rho(1-\alpha)} + b(EN_t)^{\rho\beta}(NN_t)^{\rho(1-\beta)}]^{1/\rho}$$

$$t = 1, \ldots, T$$

At its top level, this nested function has two terms. The first may be interpreted as a value-added aggregate of capital and labor, the second as an energy aggregate of electric and nonelectric energy. Both aggregates are based on a unitary elasticity of substitution between the productive factors entering at the bottom level. The parameter α (also known as kpvs) may be interpreted as the optimal value share of capital in the value added aggregate. Similarly, β (also known as elvs) may be interpreted as the optimal value share of electricity in the energy aggregate. The exponent ρ is related to ESUB (the elasticity of substitution between the energy and the value-added aggregates) through the following equation: $\rho = 1 - (1/\text{ESUB})$. (For the concepts and terminology of macroeconomic production functions and neoclassical growth theory, see Allen 1968.)

A base-year benchmarking procedure is employed to determine the coefficients a and b in each region's new vintage production function, NEWPROD_t. Let PNREF denote the reference price of nonelectric energy in the base year. (If PNREF is less than the actual base year price, there will automatically be some costless conservation in addition to that induced by the AEEI factor. This type of conservation is sometimes described as an initial momentum effect.) Neglecting the time subscripts for the base year, a first-order optimality condition implies that

$$\partial Y/\partial N = Y^{1-\rho} \, E^{\rho\beta} \, N^{\rho(1-\beta)-1}(1 - \beta)b = \text{PNREF}.$$

Except for b, each element in the preceding equation is known from the base year statistics or other input parameters. After solving for

Box 7.2

Alternative Forms of the Nested Constant Elasticity of Substitution Production Function

ETA-MACRO is based on a specific form of the nested constant elasticity of substitution (CES) production function. At the lower level, capital and labor (K and L) are nested against each other with a unitary elasticity of substitution. Capital and labor are then combined with energy (E) through the ESUB parameter. With the elasticity values that are typically employed in our model, all three inputs are said to be gross substitutes. That is, suppose that output is to be held constant and that the prices of capital and labor are also held constant. A rise in the price of energy will lead to a decline in the optimal input of energy and a rise in both capital and labor inputs.

Suppose instead that we were to adopt the functional form proposed by Burniaux et al. (1991) for their GREEN (GeneRal Equilibrium ENvironmental) model. Capital and energy are nested against each other with a low elasticity of substitution, and this aggregate is combined with labor using a higher elasticity of substitution. For output to remain constant, a rise in the price of energy will lead to a decline in the inputs of energy and capital and an increase in the input of labor. Capital and energy are then said to be complements.

Quantitatively, how important are these effects likely to be? To explore this issue, we begin with a benchmark equilibrium in which the units of measurement are chosen so that the prices of energy, capital, and labor are all unity. The leftmost column of table 7.1 shows the stylized facts adopted to characterize this case. The input-output ratios (the value shares) are 5 percent, 25 percent and 70 percent for E, K, and L, respectively. The returns to capital and labor (GDP) are 95 percent of gross output. Numerical solutions have been obtained through the MPS/GE software developed by Rutherford (1989).

Now suppose that there is a doubling of energy prices and that the quantities of K and L remain constant. Assuming a sufficiently long horizon so that the inputs of energy and the aggregate output are optimally adapted to the new price regime, what will be the effect on the GDP, the demand for energy, and the prices of capital and labor? With the elasticity parameters typically employed in ETA-MACRO and in GREEN, the GDP and the energy demand impacts are roughly comparable for both forms of production function. A doubling of energy prices leads to about a 5 percent decline in GDP and a 25 percent decline in the input of energy. The implied price elasticity of demand for energy is close to $-.4$ in both cases.

The most striking difference occurs with respect to the values of capital and labor. With the GREEN production function, there is a drastic drop in the price of capital and a modest decline in that of labor. By contrast, with the nesting adopted in ETA-MACRO, both prices decrease by the identical percentage: 5.5 percent.

In the short run, it is clear that energy usage is tied to the existing vintages of capital goods. Energy and capital are complementary, and this idea is incorporated in the putty-clay structure of ETA-MACRO. For the longer run, it is not so obvious that capital and energy inputs need to be complementary. For example, one can reduce heat losses by investing additional capital in insulation. Given enough time to adapt the structure of production to higher energy prices, there is no a priori technological reason to reject the view that energy and capital are substitutes.

ETA-MACRO is built on the assumption of Harrod-neutral technical change. This means that there is a constant long-run ratio of the quantity and the price of capital relative to that of labor (expressed in efficiency units). Alternatively, suppose that the model had been based on long-run energy-capital complementarity. If there were an upward trend in the price of energy, we would then have had to discard two of the most famous ratios in empirical economics. These are pillars of neoclassical growth theory. In view of all the uncertainties concerning the price and nonprice energy conservation parameters, we have been reluctant to drop the notion of Harrod neutrality. This is the reason for the specific form of nesting adopted in ETA-MACRO.

Table 7.1
A comparison between alternative nested CES production functions

Benchmark	GREEN F((K,E),L)	ETA-MACRO G((K,L),E)
Elasticities		
K/E:L	0.6	
K:E	0.3	
K/L:E		0.4
K:L		1.0
Output 1.000	0.985	0.982
GDP 0.950	0.906	0.907
Prices		
E 1.000	2.000	2.000
K 1.000	0.896	0.955
L 1.000	0.975	0.955
Quantities		
E 0.050	0.039	0.037
K 0.250	0.250	0.250
L 0.700	0.700	0.700

b, we employ the base year values in the production function NEW-PROD. The base year labor force index is defined as 1.0. This permits us to solve the following equation for the parameter a:

$$Y^\rho = aK^{\alpha\rho} + bE^{\rho\beta}\, N^{\rho(1-\beta)}.$$

To define the production process for new capital stocks, it is convenient to assume a geometric decay model, one in which the identical depreciation rates apply to the inputs of capital, labor and energy and also to the aggregate output from a given vintage. If the annual depreciation rate is depr and there are ten years between successive time periods, the output from old capital surviving from period t to period $t+1$ is $Y_t(1-\text{depr})^{10}$. The equation TOTALPROD$_{t+1}$ indicates that total production in period $t+1$ is the sum of the amounts produced by new investment plus the amounts produced from earlier vintages:

$$\text{TOTALPROD}_{t+1}\text{: } YN_{t+1} = Y_{t+1} - Y_t(1-\text{depr})^{10} \qquad t = 0, \ldots, T-1$$

Similarly, the equation TOTALCAP$_{t+1}$ indicates the total capital stock surviving from one period to the next:

$$\text{TOTALCAP}_{t+1}\text{: } KN_{t+1} = K_{t+1} - K_t(1-\text{depr})^{10} \qquad t = 0, \ldots, T-1$$

The labor force (measured in efficiency units) is an exogenously specified index number, L_t. Its values are $L_0 = 1$ and $L_{t+1} = (1 + \text{grow}_t)^{10}L_t$. The labor force available for new vintage production is therefore:

$$LN_{t+1} = L_{t+1} - L_t(1-\text{depr})^{10} \qquad t = 0, \ldots, T-1$$

The new vintage demands for electric and nonelectric energy are:

$$\text{NEWELEC}_{t+1}\text{: } EN_{t+1} = E_{t+1} - E_t(1-\text{depr})^{10} \qquad t = 0, \ldots, T-1$$

$$\text{NEWNON}_{t+1}\text{: } NN_{t+1} = N_{t+1} - N_t(1-\text{depr})^{10} \qquad t = 0, \ldots, T-1$$

ETA Submodel

Within ETA, the first constraints are those labeled SUPELEC. These ensure that the supplies of electric energy will be adequate to cover the demands. In period t, the decision variables $PE_{i,t}$ represent the quantity of electrical energy to be supplied by individual technologies, and the decision variables E_t indicate the quantity demanded, before allowing for the effects of autonomous energy efficiency improvements. The annual AEEI factor (compounded over successive ten-year periods) provides for a costless reduction in the amount of electrical energy to be supplied:

$$\text{SUPELEC}_t: \sum_{i \in ET} PE_{i,t} \geq E_t \prod_{\tau=0}^{t-1} (1 - AEEI_\tau)^{10} \qquad t = 1, \ldots, T$$

In the constraints governing the supply-demand balance for non-electric energy, a similar set of conventions is employed to handle autonomous energy efficiency improvements. Oil and natural gas are distinguished from each other but are viewed as perfect calorific substitutes in the SUPNON constraints. To allow for the demand and supply of natural gas, three classes of gas-related decision variables appear in these constraints: GN (consumption of gas in the form of nonelectric energy), PN_{gas-lc} (production of low-cost gas), and PN_{gas-hc} (production of high-cost gas). There is also a term to allow for the consumption of fuel oil in the electric power sector. This is why the heat rate coefficient ($htrt_{oil-r}$) is multiplied by the electric energy produced by oil-fired plants ($PE_{oil-r,t}$).

$$\text{SUPNON}_t: \sum_{i \in NT} PN_{i,t} + GN_t - PN_{gas-lc,t} - PN_{gas-hc,t} \geq$$

$$(htrt_{oil-r})(PE_{oil-r,t}) + N_t \prod_{\tau=0}^{t-1} (1 - AEEI_\tau)^{10} \qquad t = 1, \ldots, T$$

The SUPGAS constraints govern the supply-demand balance for natural gas. Neglecting interregional gas trade, the only quantities available are those supplied by low- and high-cost domestic production. The demands include those for existing gas-fired electricity plants, new gas-fired plants, and the amounts of gas employed to supply nonelectric energy demands:

$$\text{SUPGAS}_t: \text{PN}_{\text{gas-lc},t} + \text{PN}_{\text{gas-hc},t} \geq (\text{htrt}_{\text{gas-r}})(\text{PE}_{\text{gas-r},t}) + (\text{htrt}_{\text{gas-n}})$$

$$(\text{PE}_{\text{gas-n},t}) + \text{GN}_t \quad t = 1, \ldots, T$$

To avoid excessively rapid expansion of new technologies or rapid decline of obsolete ones, there are expansion or decline rate constraints of the following form. The supply expansion limits are not rigid upper bounds but are soft constraints. Growth may be accelerated but at a rising marginal cost determined by the level of the above-normal expansion variables XPE and XPN. These activities are valued not only because they enable an increase in current output but also because they provide a base for future expansion.

$$\text{EXP}_{t+1,i}: \text{PE}_{i,t+1} \leq \text{PE}_{i,t}(\text{expf})^{10} + \text{XPE}_{i,t+1} \qquad i \in \text{EX and ecap}_{t+1,i} > 0$$

$$\text{NXP}_{t+1,j}: \text{PN}_{j,t+1} \leq \text{PN}_{j,t}(\text{nxpf})^{10} + \text{XPN}_{j,t+1} \qquad j \in \text{NX and ncap}_{t+1,j} > 0$$

$$\text{DEC}_{t+1,k}: \text{PE}_{k,t+1} \geq \text{PE}_{k,t}(\text{decf})^{10} \qquad\qquad k \in \text{DL}$$

To describe the production of exhaustible resources, there is a set X that includes four categories of hydrocarbons: low- and high-cost oil, low- and high-cost gas. Proved reserves are depleted by current production and augmented by new discoveries out of the remaining stock of undiscovered resources.

In algebraic terms, these ideas are expressed as follows. For each exhaustible resource belonging to the set X, each period's current production of these resources is determined by a fixed ratio of current production to proved reserves, prv_X:

$$\text{PRVLIM}_{X,t}: \text{PN}_{X,t} = \text{prv}_X \text{RSV}_{X,t} \qquad t = 0, \ldots, T$$

Proved reserves available at time t+1 are determined by those at time t plus the time-weighted annual average of gross reserve additions less production ($RA_X - PN_X$):

$$RSVAV_{X,t+1}: \quad RSV_{X,t+1} = RSV_{X,t} + 5[(RA_{X,t+1} - PN_{X,t+1})$$

$$+ (RA_{X,t} - PN_{X,t})] \qquad t = 0, \dots, T - 1$$

Undiscovered resources remaining available at time t+1 are determined by those at time t less the time-weighted average of gross additions to proved reserves:

$$RSCAV_{X,t+1}: \quad RSC_{X,t+1} = RSC_{X,t} - 5(RA_{X,t+1} + RA_{X,t}) \qquad t = 0, \dots, T - 1$$

This is almost but not quite a constant ratio model of exhaustible resource depletion. The only element of flexibility lies in the ability to defer reserve additions. Instead of a constant ratio of reserve additions to undiscovered resources, there is an upper limit determined by the resource depletion factor rdf_X:

$$RDFLIM_{X,t}: \quad RA_{X,t} \leq rdf_X \, RSC_{X,t} \qquad t = 0, \dots, T$$

With this formulation, Global 2100 is able to incorporate the Hotelling feature of forward-looking resource depletion policies. Economic rents on exhaustible resources cannot rise more rapidly than the marginal productivity of capital. At the same time, the model is capable of representing an important real-world phenomenon. During a given year, a region may be importing oil and also engaging in domestic production out of both low- and high-cost resources. By contrast, a typical Hotelling model would rule out the possibility of simultaneous production from different resource cost categories.

The ANC constraints refer to the annual carbon emissions limits, expressed in the form of carbon emissions coefficients ($cece_i$ and $cecn_j$ for electric and nonelectric energy, respectively) multiplied by activity levels. The variables DC_t provide an option for delaying the use of carbon rights; $CARM_t$ and $CARX_t$ provide for international trade in these rights.

$$ANC_t: \sum_{i \in ET} (cece_i)(PE_{i,t}) + \sum_{j \in NT} (cecn_j)(PN_{j,t}) + DC_t - DC_{t-1} - CARM_t$$

$$+ CARX_t \leq nenc + carlim_t \qquad t = 0, \ldots, T$$

The right-hand-side constant $carlim_t$ is the carbon limit applicable to the specific region, and nenc represents an adjustment to allow for nonenergy uses of carbon-based fuels (such as lubricating oils and asphalt).

The dual variables for the ANC_t constraints represent the present value of carbon rights available in period t. To convert into future values, these are divided by the dual variables for the CC_t constraints. Alternatively, they may be divided by the dual variables of the $COSTNRG_t$ constraints. Since the variable EC_t enters only into the $COSTNRG_t$ and the CC_t constraints, the dual variables of these two constraints are identical.

The carbon consumption deferral activities DC_t play a key role when the value of carbon rights is rising rapidly. Since no upper bounds are imposed on their level, they ensure that the value of carbon cannot rise at a more rapid rate than that determined by the marginal productivity of capital. These variables provide for the endogenous allocation of carbon rights between successive periods, but the allocation is subject to the constraint that cumulative emissions cannot exceed an exogenously specified total. This feature of Global 2100 resembles the Hotelling model of exhaustible resources. A look-ahead analysis leads to smoother price trajectories than are characteristic of recursive myopic simulations.

The final constraints are those that provide for feedback from ETA to the MACRO submodel. The variable EC_t represents the total of all costs related to the energy supply sector. In this sum, the first components are determined by the individual cost coefficients ($ecst_i$ and $ncst_{t,j}$) and the activity levels ($PE_{i,t}$ and $PN_{j,t}$) for the supply of electric and nonelectric energy, respectively. An additional term is included to allow for the fact that oil and natural gas are imperfect economic substitutes. The GN_t activity describes the nonelectric uses of natural gas, and the ogdp coefficient reflects the historic oil-gas price differential for this activity.

There are quadratic penalty terms associated with the above-normal expansion activities $XPE_{i,t}$ and $XPN_{j,t}$. Each penalty coefficient is chosen so that the marginal cost of energy production is twice its long-run normal level if an expansion activity is pushed to the maximum feasible rate during the first decade in which the specific technology becomes available. For an example of how these expansion limits interact with energy prices, recall figure 3.8. Over the long run, the marginal cost of energy is determined by the backstop technology. During the transition period, however, the expansion constraints lead to a period of overshoot above the backstop level. These effects are moderated but not eliminated by the operation of the above-normal capacity expansion activities.

The carbon import and export activity levels are employed only in scenarios for which there is international trade in carbon rights. These trade activities are associated with two coefficients in the energy cost equation: the international carbon price, $carp_t$, and a transaction cost parameter mxdif, measuring the maximum difference between import and export prices. This term ensures that trade will occur only when there is a significant interregional differential in carbon prices.

$COSTNRG_t$:

$$1000\,EC_t = \sum_{i \in ET}(ecst_i)(PE_{i,t}) + \sum_{j \in NT}(ncst_{t,j})(PN_{j,t})$$

$$+ (ogdp)(GN_t)$$

$$+ .5\sum_{i \in EX}\frac{(ecst_i)}{ecapfy_i}(XPE_{i,t})^2 + .5\sum_{j \in NX}\frac{(ncst_{t,j})}{ncapfy_j}(XPN_{j,t})^2$$

$$+ carp_t(CARM_t - CARX_t) + .5\,mxdif\,(CARM_t + CARX_t)$$

$$t = 1, \ldots, T$$

International Trade in Oil and Carbon Rights

In the preceding formulation, each of the five regions is treated in isolation from the others, and the analysis proceeds by parallel independent computations. For crude oil, the model is almost consistent. That is, the price trajectory is determined exogenously by OPEC. The

OECD nations are viewed as price takers. Upper bounds are imposed on oil exports from the Soviet Union and China, and the ROW region is the residual supplier. If it is not optimal for the ROW region to supply this quantity, there is an excess demand gap at a specific point in time, a clear indication that OPEC has set too low a price on its oil exports. Through trial and error, it is a relatively simple matter to adjust the OPEC price trajectory upward so as to avoid an excess demand gap for crude oil.

If we wish to introduce the possibility of international trade in carbon rights, trial and error is not nearly so easy. This is why we have turned to the decomposition algorithm described in chapter 8. During this procedure, we repeatedly make use of parallel computations for each of the regions. For short, we employ the acronym 5R to refer to the five independent nonlinear optimizations. Each is of the same general structure as the single-region model formulated in this chapter.

Box 7.3

Decisions under Uncertainty

In order to allow for decisions under uncertainty (as in chapter 4), we attach a state-of-the-world subscript to each decision variable and to each constraint in Global 2100. Denote the alternative states as $s = 1, \ldots, S$. For example, if we are uncertain as to which of three carbon limits will characterize the future, this means that $S = 3$. There are three times as many variables and more than three times as many constraints as in the deterministic version of the model. With this increase in size, the computing time grows from 2 minutes to 15 minutes on a 25 MHz 486 desktop computer. Altogether, there are about 1300 constraints and 1800 decision variables when $S = 3$.

In this two-stage problem, the original decision variables may be grouped into two categories: those that must be determined prior to the resolution of uncertainty and those that must be determined afterward. Let the vector x_{1s} denote the first group of variables and x_{2s} the second group. In the specific case described in chapter 4, all uncertainties are resolved between 2010 and 2020. The vector x_{1s} therefore refers to decisions that must be taken during the initial time periods (2000 and 2010). The vector x_{2s} refers to decisions taken during the second stage (2020, 2030, ..., 2100). Each of the decision variables may be subject to individual upper or lower bounds.

Let p_s denote the probability of state-of-world s. Let $f_s(x_{1s}, x_{2s})$, a concave nonlinear function, denote the discounted utility of consumption in state-of-world s. The values of the decision variables are chosen so as to maximize the expected discounted utility of consumption. The objective function is therefore:

$$\text{maximize}_{x_{1s}, x_{2s}} \quad \sum_{s=1}^{S} P_s \, f_s(x_{1s}, x_{2s}).$$

There are constraints for each state-of-world s. The constraint sets are convex. They may be written:

$$g_s(x_{1s}, x_{2s}) \leq 0 \quad (s = 1, \ldots, S).$$

Without any further restrictions, our model would describe the situation of "learn–then act". In order to represent act–then learn, we add equations ensuring that the initial decision variables take on the identical values, regardless of the eventual state-of-world—that is:

$$x_{1s} = x_{11} \quad (s = 2, \ldots, S).$$

It is straightforward to formulate a two-stage uncertainty model and to obtain numerical solutions when S = 3. Numerical solutions may be difficult, however, when S is large. For solving large-scale linear programming problems of this type, see Dantzig and Glynn (1990).

References

R. G. D. Allen. 1968. *Macroeconomic Theory*. Macmillan, New York.

A. Brooke, D. Kendrick, and A. Meeraus. 1988. *GAMS: A User's Guide.*, Scientific Press, Redwood City, Calif.

J. M. Burniaux, J. P. Martin, G. Nicoletti, and J. Oliveira Martins. 1991. "GREEN . . . A Multi-Region Dynamic General Equilibrium Model for Quantifying the Costs of Curbing CO_2 Emissions: A Technical Manual," Economics and Statistics Department, Organisation for Economic Co-operation and Development, Paris.

S. Chakravarty. 1969. *Capital and Development Planning*. MIT Press, Cambridge, Mass.

G. Dantzig and P. Glynn. 1990. "Parallel Processors for Planning under Uncertainty," *Annals of Operations Research* 22, pp. 121.

J. Edmonds and J. Reilly. 1985. *Global Energy: Assessing the Future*. Oxford University Press, New York.

T. Rutherford. 1989. "General Equilibrium Modelling with MPS/GE," Department of Economics, University of Western Ontario, April.

R. M. Solow. 1974. "The Economics of Resources or the Resources of Economics," *American Economic Review* 64 (2), May.

A. Svoronos. 1985. *Duality Theory and Finite Horizon Approximations for Discrete Time Infinite Horizon Convex Programs.* Department of Operations Research, Stanford University, April.

8 A Decomposition Procedure for International Trade in Carbon Emission Rights

Introduction

In the preceding chapter, Global 2100 was described as though all computations were performed in parallel for the five geopolitical regions. Except for oil trade, these regions were treated independently. Here we describe a procedure that helps to quantify the potential for international trade in carbon permits. Each individual region is viewed as a price taker and as a possible importer or exporter of carbon rights. Each is coupled to the others through the international price of these rights.

Since this is an intertemporal problem, the time path of prices must be determined so as to equilibrate supplies and demands during each period simultaneously. We can no longer formulate the overall problem as 5R—five independent nonlinear optimizations. Instead, we must solve an equilibrium problem in which all choices are integrated through an international market in carbon rights. Each of the five agents has an independent objective function and resource endowments. For a masterful review of such computations, see Scarf 1984.

An equilibrium problem is easy to formulate, but this one suffers severely from the curse of dimensionality. Since each commodity is differentiated between regions and time periods, there are approximately 5000 prices and quantities are to be determined. A problem of this size and structure cannot be solved with any of the general-purpose algorithms currently available (Rutherford 1989). As a practical alternative, we have turned to a decomposition procedure. There are heuristic elements in the method to be described, and the results do not have a high degree of numerical precision. These shortcomings are real, but at some point within the near future it should be possible to remedy them.

The decomposition principle was originally applied to linear programming by Dantzig and Wolfe (1961). They showed how a large model might be decomposed into a master and one or more subproblems—each a low-dimensional independent linear program. Iterations proceed by sending price signals from the master to the subproblems and quantity signals in the reverse direction. In the final iteration, there may also be quantity signals sent from the master to the subproblems, but these allocations are compatible with decentralized optimization. They are needed only to resolve the indeterminacies associated with constant returns to scale. That is, under constant returns, prices do not uniquely determine the quantities of inputs and outputs.

Several of these ideas were adapted to an equilibrium model of international trade by Mansur and Whalley (1982). Their paper applies to pure exchange models, and they do not discuss the application of these procedures to economies with production. Like Dantzig and Wolfe, they draw a sharp distinction between tradables (prices and quantities determined by the master problem) and nontradables (prices and quantities determined within the individual subproblems). Mansur and Whalley have summarized their algorithm in this way: "The method involves the generation of labels for vertices on a master simplex through the separate solution of sub-equilibrium problems whose parameters are determined by the coordinates of the vertex on the master simplex" (p. 154). They provide small-scale numerical examples indicating that their decomposition method has a significant potential for computational improvements over the usual algorithms that ignore the distinction between tradables and nontraded goods.

This chapter reports numerical results employing a procedure motivated partly by Dantzig-Wolfe and partly by Mansur-Whalley. There is price-guided decentralization, there are quantity responses from the subproblems, and there are quantity allocations of tradable goods at the final iteration. A solution requires about 4 hours on a 25 MHz 486 desktop computer. There is no proof that the decomposition method will converge under all circumstances. This issue is left as a challenge to subsequent investigators.

Notation and an Outline of the Decomposition Procedure

Figure 8.1 provides an outline of the procedure employed here. Recall that 5R is the acronym describing the parallel five-region optimizations. These are performed in three modes: no trade in carbon rights,

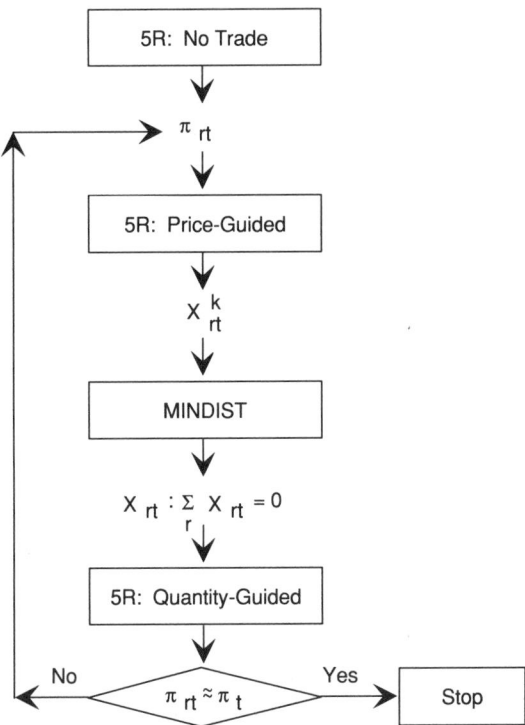

Figure 8.1 Decomposition procedure

price-guided decentralization, and quantity-guided decentralization. To solve the master problem, we employ a minimum distance convex combination of the proposals received from the subproblems.

The master problem is an unconventional use of linear programming. It resembles the use of a linear programming model to minimize the sum of the absolute errors in the statistical estimation of parameters subject to linear constraints. There are also instructive points of resemblance between the minimum distance formulation and the Scarf (1984) vector labeling technique.

In this specific application of the decomposition procedure, the only tradables consist of carbon rights and a numeraire good produced in each of the regions. Let π_{rt} denote the price of carbon rights in region r during period t. In equilibrium, this price will be identical in all regions; that is, $\pi_{rt} = \pi_t$ (for all t).

Let x_{rt} denote the net exports of carbon rights from region r during period t. Note that x_{rt} may be positive, negative, or zero. In equilibrium, $\sum_r x_{rt} = 0$ (for all t). That is, the international market for carbon rights must clear during each period t. Our decomposition procedure is designed so that the quantity equilibrium conditions are met exactly (subject to certain qualifications discussed below). The price equilibrium conditions are met only approximately. If there are significant discrepancies in these prices, one cycles back to the second step of the algorithm and eventually arrives at an equilibrium allocation of carbon rights. The equilibrium is characterized by (approximately) equal values of these rights in all regions.

Numerical Results for 5R:
No Trade and Price-Guided Decentralization

According to figure 8.1, the first step in the decomposition procedure is to determine the carbon prices that would prevail within each individual region in the absence of trade. That is, we set $x_{rt} = 0$ and solve 5R for the prices π_{rt}. We have already seen a typical set of numerical results in figure 5.6.

Despite the differences from one region to another, there is a general pattern to the price trajectories. During the initial years, it is frequently optimal to delay the use of carbon rights (that is, to set the variables $DC_t > 0$). Prices then rise at the same rate as the marginal productivity of capital in each region—about 5 percent annually. During the later years, carbon prices stabilize at a level determined by the carbon coefficient and the cost difference between a carbon-free and a carbon-based backstop. (Except for the Soviet Union, this long-run level is $208 per ton.) For most of the regions, there is an intermediate period during which the new technologies are limited in their rate of deployment, and the value of carbon therefore overshoots the backstop level.

It takes no more than a back-of-the-envelope calculation to determine the carbon price during the backstop phase. If all regions have identical energy supply options at that time, there is no motivation for trade. Each region can be self-sufficient in carbon rights. Alternatively, if there are systematic differences among regions, the carbon price will be determined by the least-cost region, and it will export carbon rights to the others. There is no need to construct an intertemporal equilibrium model in order to analyze international trade in carbon rights.

An intertemporal model is useful only during the transition period, when prices are first rising and then falling toward the backstop level. Moreover, an intertemporal model is needed in order to indicate the beginning date for the backstop phase.

A major simplification is suggested by the no-trade case. Figure 5.6 suggests that we can truncate the analysis of international trade in 2050. By that date, virtually all the regions have reached the backstop era.

Now return to figure 8.1. For each region r, the no-trade price vectors π_{rt} are internally consistent. That is, carbon prices do not rise more rapidly than the marginal productivity of capital. These price vectors are applied in succession to solve 5R for the case of price-guided decentralization, allowing for positive or negative values of the net export variables x_{rt}. Thus we first solve 5R using $\pi_{USA,t}$, the price vector that emerged from solving the United States submodel under the assumption of no trade. Similarly, we solve 5R using the price vectors for the other four regions (other OECD,..., rest of world): $\pi_{OOECD,t}, \cdots, \pi_{ROW,t}$.

In general, we solve the price-guided version of 5R for each of K carbon price vectors and record the regional distribution of net exports as x_{rt}^k (for $k = 1,..., K$). In the first pass through the decomposition procedure, there are only five carbon price vectors, and K = 5. During each subsequent pass, the price-guided 5R calculations lead to five additional net export vectors. These results are used as an input to the quantity-guided calculations, and five additional price vectors are generated. In successive passes, K grows from 5 to 10 to 15, and so on.

The Master Problem MINDIST

The principal inputs to the master problem are the net export records associated with each of the K price vectors. For a given price vector k, the price-guided version of 5R determines $\bar{x}_t^k = \sum_r x_{rt}^k$ equals the global level of net carbon exports during period t. In general, the kth price vector will not lead to a global equilibrium in net export quantities; that is, $\bar{x}_t^k \neq 0$. The master problem is formulated, however, so that a weighted average of these net export vectors will add up to zero. The weights are chosen so as to minimize the weighted distance from zero net export demands.

This criterion leads to the name MINDIST as the acronym for the master problem. It is a low-dimensional linear program. The decision variables are all nonnegative. There are weights on the K individual net export records. There are artificial high-cost error vectors to ensure a feasible solution. There are also carbon emissions delay activities that make it possible to hold over carbon rights from one period to the next. The delay activities duplicate those already included in each regional subproblem. The list of decision variables is as follows:

λ_k = weight to be attached to the net export vector $\bar{x}_t^k$
PER_t = positive error, period t
NER_t = negative error, period t
DC_t = carbon delay activities to allow for carry-over from period t to t + 1

Within MINDIST, the minimand is the sum of two terms: the true objective function (the weighted sum of the distances from zero net export demands) and a penalty cost associated with the positive and negative errors:

$$\text{minimand} = \sum_k [\sum_t (\bar{x}_t^k)^2]\lambda_k + M \sum_t [PER_t + NER_t]$$

where the coefficient M is given a high enough value so as to discourage positive intensities for the error activities. For the example described in this chapter, the optimal value of one of the error variables was positive during the first cycle of MINDIST calculations but zero thereafter. If PER_t (NER_t) is positive, this is an indication that the K original price vectors do not provide a sufficiently rich sample of candidates for an equilibrium solution. We then raise (lower) the tth component of one of the price vectors by an arbitrary amount, rerun the price-guided version of 5R, and insert a new vector of net exports into MINDIST.

In this example, the number of trading regions is nearly as large as the number of (dated) commodities to be traded. Had the number of traded commodities been much larger and the number of regions remained the same, it might have been difficult to arrive at a good choice for the arbitrary changes in the original price vectors derived from the case of no trade.

When the intertemporal equilibrium problem contains T time periods, there are T+2 constraints in this linear program. The first one is a statement that the nonnegative weights must add to unity:

$$\sum_k \lambda_k = 1$$

The next T constraints represent the global equilibrium condition that the weighted average of net exports must add to zero but that due allowance be made for the error variables and for the option of delaying the use of carbon rights from period t to period t+1:

$$\sum_k \bar{x}_t^k \lambda_k - DC_t + DC_{t-1} + PER_t - NER_t = 0 \qquad (t = 1, \dots, T)$$

The final constraint is a statement that no carbon rights are to be carried forward beyond the planning horizon of MINDIST (not beyond 2050):

$$DC_T = 0$$

Clearly it is feasible to introduce the carbon delay variables DC_t into the master problem. They are duplicates of identical activities that are already available within each regional subproblem. They lead, however, to an element of indeterminacy. It makes no difference to which region we attribute their operation.

Typically, it is optimal to assign positive values to one or more of the DC_t variables. This can be interpreted as though there were a global bank for the accumulation of carbon rights, and unused credits in period t may be carried forward for use in t+1. This is a case in which there is no conflict between environmental and economic goals. The DC_t activities help to improve overall economic efficiency and at the same time help to delay the accumulation of greenhouse gases.

Numerical Results for 5R: Quantity-Guided Decentralization

As in the final step of Dantzig-Wolfe decomposition, here too there is a final step in which quantity guidelines are sent from the master problem to the subproblems. (Recall the bottom box in figure 8.1.) These quantities are based on the net export values derived from the price-guided phase of 5R and also on the weights derived from the optimal solution to MINDIST. They are calculated as follows:

$$x_{rt} = \sum_k x_{rt}^k \lambda_k$$

If zero values are assigned to the error variables and to the delay variables (PER_t, NER_t, and DC_t), the quantity assignments will satisfy the global equilibrium condition that net exports be zero, that is, $\sum_r x_{rt} = 0$ (for $t = 1, \ldots, T$). If the delay variables are operated at positive levels in some periods, it is straightforward to show that cumulative net exports are nonnegative in each period, and that they are zero for the T periods as a whole. The global quantity equilibrium conditions are satisfied as follows:

$$\sum_{\tau=1}^{t} \sum_r x_{r\tau} \geq 0 \quad (t = 1, \ldots, T - 1) \quad \text{and} \quad \sum_{\tau=1}^{T} \sum_r x_{r\tau} = 0$$

With the net export allocations x_{rt}, we run the quantity-guided version of 5R (the bottom box in figure 8.1), which leads to a new set of regional carbon prices π_{rt}. With a bit of luck, these will be close to each other, and we can average them to arrive at our final goal: the international equilibrium prices, π_t. After just four cycles (K = 20), we obtain the results shown in figure 8.2. This refers to carbon stabilization scenario I described in chapter 5. It provides graphical confirmation that the decomposition procedure leads to regional prices that match each other reasonably well in most periods; that is, $\pi_{rt} \approx \pi_t$.

Concluding Comments

There are weaknesses in the decomposition procedure outlined here, for there is no proof that a satisfactory answer will always be obtained. It is possible that superior results would be produced by other methods such as Negishi weights, as Ginsburgh and Van der Heyden (1985) have reported.

There is a role for several methods of computation, and there is no reason to allow the best to be the enemy of the good. The decomposition methodology provides useful insights into the benefits from trade in carbon rights, and it serves as an instructive starting point for the next generation of models in this area.

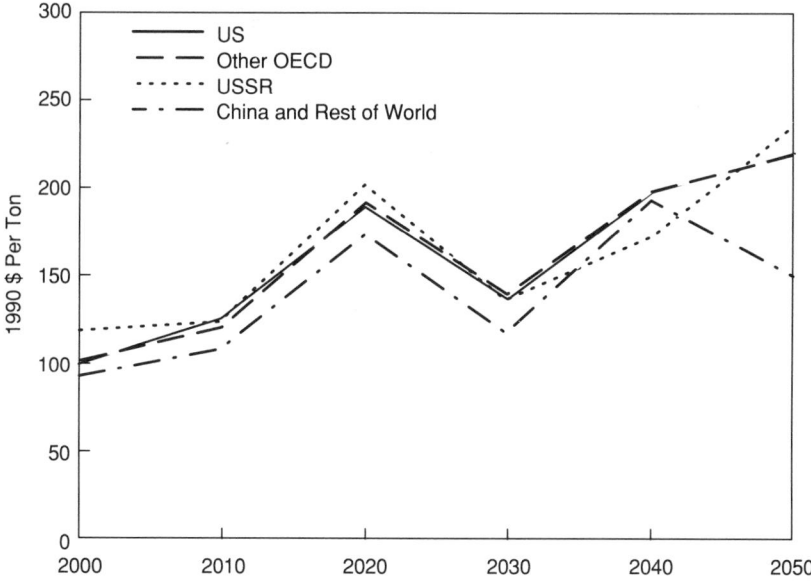

Figure 8.2   Carbon tax rates: stabilization scenario I

References

G. Dantzig and P. Wolfe. 1961. "The Decomposition Algorithm for Linear Programs." *Econometrica* 29.

V. Ginsburgh and L. Van der Heyden. 1985. "General Equilibrium with Wage Rigidities." *Mathematical Programming Study* 23, October.

A. Mansur and J. Whalley. 1982. "A Decomposition Algorithm for General Equilibrium Computation with Application to International Trade Models." *Econometrica* 50, November.

T. Rutherford. 1989. *General Equilibrium Modelling with MPS/GE.* Department of Economics, University of Western Ontario, April.

H. Scarf. 1984. "The Computation of Equilibrium Prices." In H. Scarf and J. Shoven (eds.), *Applied General Equilibrium Analysis.* Cambridge University Press, Cambridge.

9 Estimating the Energy Conservation Parameters: An Experiment in Backcasting

Energy Conservation Parameters

Energy consumption need not grow at the same rate as the GDP. Over the long run, they may be decoupled. In Global 2100, these possibilities are summarized through two macroeconomic parameters for energy conservation: ESUB (the elasticity of price-induced substitution) and AEEI (autonomous energy efficiency improvements). (For details on how these enter the macroeconomic production function, see the equation NEWPROD$_t$ described in chapter 7.)

If there is sufficient time for the adaptation of capital stocks, most analysts will agree that there is a good deal of possible substitutability between the inputs of capital, labor and energy. The degree of substitutability will affect the economic losses from energy scarcities and price increases. In our aggregate model, the ease or difficulty of these price-induced trade-offs is summarized by ESUB. The higher is the value of ESUB, the less expensive it is to decouple energy consumption from GDP growth during a period of rising energy prices.

When energy costs are a small fraction of total output, ESUB is approximately equal to the absolute value of the price elasticity of demand. In Global 2100, this parameter is measured at the point of secondary energy production: electricity at the busbar, crude oil and synthetic fuels at the refinery gate. For the OECD countries, our standard assumption is that ESUB = .40; that is, a 1 percent price increase will lead to a decline of 0.40 percent in the demand for energy.

In addition to the reductions in energy demand brought about by rising energy prices, there is also the impact of the AEEI. Nonprice efficiency improvements may be brought about by deliberate changes in public policy, such as mandatory efficiency labeling for appliances.

Autonomous trends may also occur as a result of shifts in the basic economic mix away from manufactured goods and toward more services.

For the OECD countries, we have assumed an AEEI of 0.5 percent annually throughout the twenty-first century. At constant energy prices, this would mean that if the GDP grows at 2.5 percent annually during the initial years, energy demands would grow at 2.0 percent. From 2050 onward, if the GDP grows at the annual rate of 1.0 percent, energy demands would grow at only 0.5 percent. Alternatively, this process can be described as a downward trend in the income elasticity of demand for energy. With these GDP growth rates and AEEI values, the income elasticity would begin with a value of 0.8 (= 2.0/2.5), and it would drop to 0.5 (= 0.5/1.0) during the second half of the twenty-first century.

The model is designed so that distinguishing between the ESUB and the AEEI parameters is straightforward. For many purposes, it may be difficult to draw a clear distinction between price and nonprice conservation. For example, energy prices are not affected directly by a gas guzzler tax or by preferential loans for energy conservation. These tax and subsidy policies operate indirectly through price signals affecting the cost of capital goods. ETA-MACRO is not intended for this type of country-specific analysis. It is a top-down model for providing a long-term global perspective.

For the most part, our estimates of ESUB and the AEEI have been chosen so that energy demands will match the conventional wisdom expressed in the median responses to the International Energy Workshop polls reported by Manne and Schrattenholzer (1989). In this chapter, we will attempt a backcasting exercise: estimating the conservation parameters for the United States during the three decades extending from 1960 through 1990. We have not gone back earlier than 1960 because otherwise we could easily confound changes in total demands with changes in the fuel mix. As recently as 1950, coal provided 38 percent of total primary energy consumption. By 1970, coal's share had dropped to 18 percent.

Just as with conventional econometrics, we see how well the energy conservation parameters can be chosen so as to match the past. For projections into the twenty-first century, we can be equally sure of two contradictory maxims: history does not repeat itself, and those who ignore history are doomed to repeat it.

Rather than a formal maximum likelihood estimation procedure, we experiment with alternative values of the ESUB and AEEI parameters to see which combination leads to the closest match with energy demands in 1990. We also check to see that the model handles interfuel substitution and electrification trends reasonably well through the electricity value share parameter ELVS.

Unlike conventional econometrics, we must face up to the complexities introduced by the perfect foresight and putty-clay features built into this intertemporal model. The input-output coefficients for each successive age cohort of energy-using equipment are chosen so as to be optimal for the discounted future trajectory of prices. Subsequent to the initial capital investment, there are no changes in these coefficients. Gross investment therefore governs the time lags of energy demands in responding to price changes. Investment depends in turn on the rates of depreciation and net economic growth.

The model is benchmarked as though inputs and outputs were optimally adjusted to the prices prevailing during the base year of 1960. With an intertemporal model based on perfect foresight, the demand for energy in 1970 is affected not only by the low prices prevailing in that year but also by the anticipation of higher prices in 1980 and 1990. Beyond 1990, for consistency with the projections reported elsewhere in this book we have assumed further increases in energy prices.

The Data and Model for 1960–1990

Table 9.1 contains the 1960–1990 data and the sources employed for the backcasting experiment. For consistency with U.S. energy statistics, the units of measurement were shifted from exajoules to quads and from $/GJ to $/MMBtu. The rates of potential GDP growth were chosen so that the realized would match with the historical values reported here. It turns out that the potential rate of GDP growth is the principal determinant of realized GDP. There are only second-order GDP effects associated with the specific choice of the AEEI and ESUB parameters.

The implicit GDP deflator has been employed so as to convert all prices into constant 1990 dollars. Two time series of energy demands are available on a consistent basis: total primary energy consumption (quadrillion Btu's) and electricity generation (trillion kWh). Nonelectric energy demands are deduced as a residual, taking the heat rate for electricity generation at its average value of 10,200 Btu per kWh.

Table 9.1
Data employed in the backcasting experiment (dollars in constant 1990 purchasing power)

Year	Quantities of energy demanded				Energy prices		
	GDP ($ trillion)	Electricity (TkWh)	Nonelectric (quads)	Total primary (quads)	Electricity at busbar ($/MkWh)	Nonelectric, refiners' acquisition cost ($/MMBtu)	Electricity's value share
1960	2.189	.756	36.04	43.75	38.34	2.11	.276
1970	3.176	1.532	50.73	66.36	27.15	1.84	.309
1980	4.190	2.286	52.64	75.96	48.26	7.43	.220
1990	5.463	2.805	52.84	81.45	39.95	3.83	.356
Annual growth rates, (%):							
1960–1970	3.8	7.3	3.5	4.3	−3.4	−1.4	
1970–1980	2.8	4.1	0.4	1.4	5.9	15.0	
1980–1990	2.7	2.1	0.0	0.7	−1.9	−6.4	

Sources: Energy Information Administration, *Annual Energy Review*, 1983, pp. 5, 127, 131, 195, 207; Energy Information Administration, *Monthly Energy Review*, March 1991, pp. 7, 84, 99, 109; Council of Economic Advisers, *Annual Report*, 1990, pp. 288, 290.

Throughout this period, oil was the marginal source of nonelectric energy. There were price controls on natural gas and environmental limits on the use of coal. Accordingly, the price of nonelectric energy is defined to be the refiners' acquisition cost of crude oil. No historical series is directly available for the busbar price of electricity; instead, it is assumed that the industrial price of electricity is a close proxy and that the busbar price lies 15 percent below this level. The difference is associated with the costs of high-voltage transmission and distribution.

For the backcasting experiment, the algebraic form is identical to that described for the MACRO constraints and the energy cost equation $COSTNRG_t$ in chapter 7. The major difference is a drastic simplification in the ETA submodel. Here there is only one technology available for supplying electricity, and there is only one for nonelectric energy. No limits are imposed on their availability, but their costs vary over time, as shown in table 9.1. In backcasting, we are concerned only with the macroeconomic submodel of energy demands, not with the process analysis submodel of energy supplies.

The time horizon begins with the base year of 1960 and includes the following years for which we have statistics: 1970, 1980, and 1990. To avoid edge effects, the horizon extends for three additional projection periods: 2000, 2010, and 2020. For the post-1990 years, the growth of demand is based on the continuation of the 1970–1990 rates of GDP growth.

The statistics in table 9.1 are consistent with most observers' impressions of the 1960–1990 period:

- Real energy prices declined during the first decade, exploded during the second, and dropped again during the third.
- There was a significant slowdown of GDP growth from the first decade to the two that followed. Among the many competing hypotheses on the causes for this slowdown are a lower rate of growth of employment, a lower rate of human capital formation, a lower rate of domestic savings, and a lower rate of physical capital formation. It is difficult to sort out the effects of these macroeconomic factors from the environmental restrictions and the rise of energy prices that occurred simultaneously during the decade 1970–1980.
- Nonelectric energy demands were closely coupled with GDP during the first decade but were decoupled during the two later decades.

Some of this decoupling may be attributed to price-induced conservation (the ESUB parameter) and some to autonomous conservation trends (the AEEI parameter). These factors offset each other during the first decade but began to reinforce each other during the second and third decades.

• There was a decline in the ratio of electric to nonelectric energy prices and an increase in the ratio of electric to nonelectric energy demands. For the three decades as a whole, there was a modest increase in electricity's value share. This would be consistent with a greater than unitary elasticity of substitution between electric and nonelectric energy. For purposes of backcasting, as well as for projections into the twenty-first century, we have assumed a unitary elasticity of interfuel substitution and set electricity's long-run value share, ELVS = .33. Table 9.2 contains a complete listing of the macroeconomic parameters employed in the backcasting experiment.

Perfect Foresight vs. Static Expectations

In Figure 9.1, the solid line indicates the evolution of TPE (total primary energy) demands between 1960 and 1990. The actual values are compared with those that would be projected with the base case parameters: ESUB = .4; AEEI = .50 percent per year. The dashed line (PF) is based on the perfect foresight feature employed in ETA-MACRO. The dotted line (SE) is based on static expectations—calculating each period's demands separately on the assumption that future prices will not change from those that prevail currently. Figure 9.2 contains a similar comparison for the actual versus projected values of electricity generation.

When we employ the base values for the energy conservation parameters, the SE and PF projections match, reasonably well with the actual values for 1990. There is only one case in which the two results are significantly different: total primary energy consumption in 1970. The difference can be attributed to anticipatory behavior during a period of rising energy prices. Energy consumers make their decisions knowing that there are rigidities in the adjustment of the capital stock. Perfect foresight leads to a low level of energy consumption prior to the oil price explosion of the 1970s.

Table 9.2
Additional parameters

Macroeconomic parameters		
GDP0	1960 GDP (1990 dollars)	$2.189 trillions
KGDP	1960 capital-GDP ratio	2.4
KPVS	Capital's value share	.24
DEPR	Annual depreciation rate	5%
ROR	Net annual rate of return on capital = (KPVS/KGDP) − DEPR	5%
Energy demand parameters		
ESUB	Elasticity of substitution between capital-labor and energy	.40 (base case)
AEEI	Autonomous energy efficiency improvement rate	.50%/year (base case)
PNREF	Reference (benchmark) price of nonelectric energy in 1960	$2.11/MMBtu
Electrification parameter		
ELVS	Electricity's value share	.33

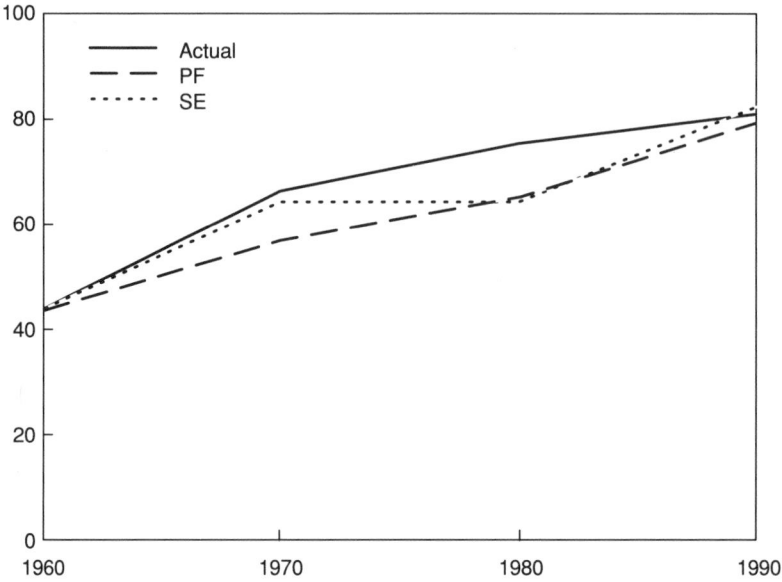

Figure 9.1 Total primary energy consumption (quads): actual vs. perfect foresight and static expectations

With static expectations, there is a striking contrast. The 1970 demand decisions are based on energy prices remaining at that year's level over the indefinite future. In this case, there is a sizable increase in the projection of 1970 energy demands. TPE consumption is almost as high as the actual value in that year.

If static expectations provide this improvement in backcasting accuracy, why don't we employ this hypothesis for projections into the twenty-first century? Why bother with the implausible assumption of perfect foresight? The answer is straightforward. If the modeler has a specific view on how energy prices are likely to evolve, static expectations imply that the modeler has more information than that available to the economic agents whose behavior is being modeled. We are skeptical of perfect foresight and even more skeptical that modelers are systematically better informed than the producers and consumers who are making economic decisions on the trade-offs between capital, labor, and energy.

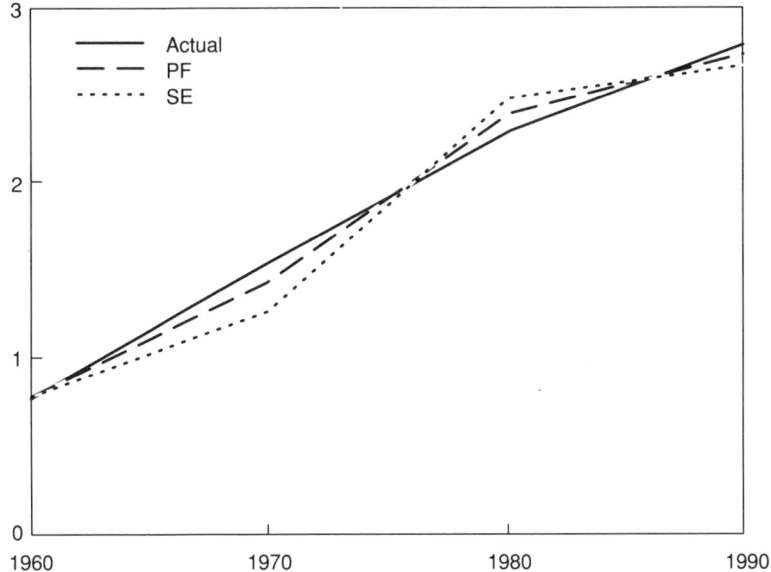

Figure 9.2 Electricity generation (TkWh): actual vs. perfect foresight and static expectations

Alternatives to the Base Case Parameter Values

Figures 9.1 and 9.2 show that the base case parameter values lead to demand projections that match reasonably well with the actual values of energy demand in 1990. What about alternative values for the ESUB and AEEI? Might they not provide an even better fit?

To check for this possibility, we have let ESUB range from 0.3 to 0.5 and varied the AEEI between 0.0 and 1.0 percent per year. Figures 9.3 and 9.4 show how the ESUB and AEEI jointly affect the ratio of projected to actual TPE and electricity demands in 1990. A perfect forecast would imply a ratio of 1.0. None of these parameter combinations leads to a perfect forecast, but the base case provides a better fit than any of the eight alternatives shown.

These calculations virtually rule out the possibility of assigning high values to both of the conservation parameters. If we adopt an AEEI value as high as 1.0 percent, we systematically underestimate the energy demands, independently of ESUB. Only by assigning a low value to ESUB (.30) can we justify an AEEI higher than 0.5 percent.

Despite anecdotal evidence on efficiency improvements in specific sectors, the aggregate results do not support the hypothesis of an AEEI

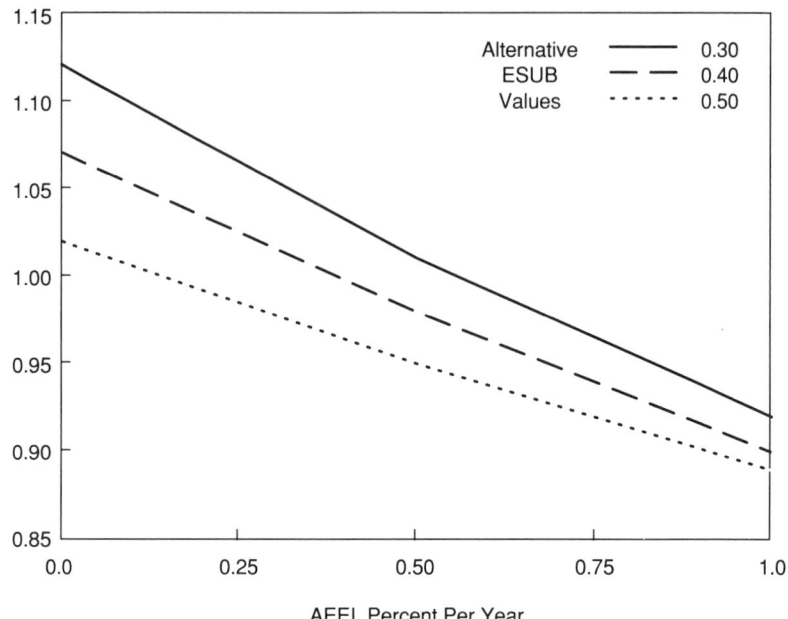

Figure 9.3 Ratios of TPE consumption: 1990 projected and actual

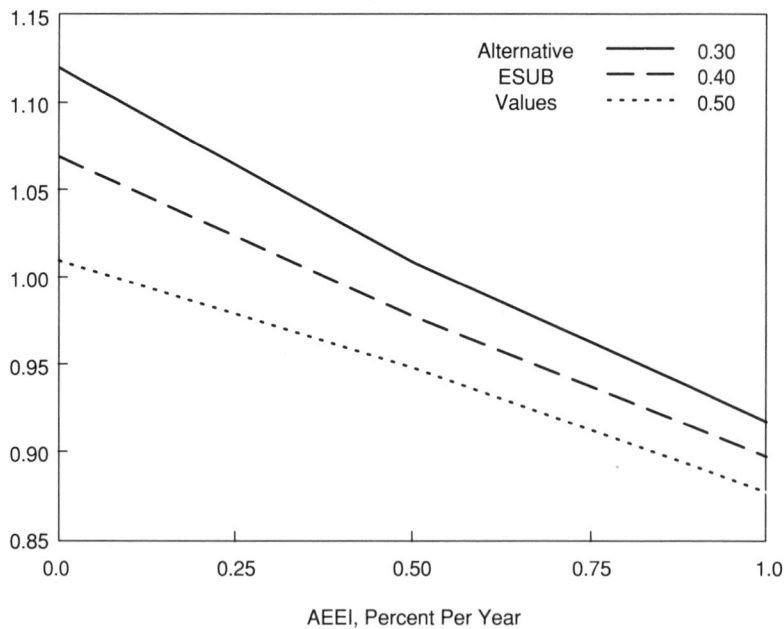

Figure 9.4 Ratios of electricity generation: 1990 projected and actual

as high as 1.0 percent. Our estimates are much more nearly in line with the formal econometric results reported by Hogan and Jorgenson (1990) than with the technological estimates reported by Williams (1990).

What are we to make of all this? The historical record is ambiguous. We cannot easily sort out how much of the post-1970 decoupling between energy consumption and GDP growth is attributable to price-induced conservation and how much to autonomous factors. In the aggregate, an AEEI value as high as 1.0 percent does not provide a good fit to the historical record. To project a 1.0 percent annual conservation trend for the twenty-first century would require the assumption of a discontinuity in the behavior of the U.S. political and economic system. This type of discontinuity could indeed occur, but it will not be effortless. All that our backcasting experiment can provide is an indication of whether a conservation trend of this magnitude is consistent with the history of the past three decades.

References

A. S. Manne and L. Schrattenholzer. 1989. "The International Energy Workshop—a Progress Report." *OPEC Review*, Winter.

W. W. Hogan and D. W. Jorgenson. 1991. "Productivity Trends and the Cost of Reducing CO_2 Emissions." *Energy Journal* 12 (1), January.

R. H. Williams. 1990. "Low-Cost Strategies for Coping with CO_2 Emission Limits." *Energy Journal* 11 (4), October.

Appendix A
Macro Assumptions
(MACRO.TAB)

* UNITS: TRILLION DOLLARS, EXAJOULES, TKWH, BILLION TONS OF CARBON.

TABLE MACRO(MP, RG) MACROECONOMIC AND OTHER PARAMETERS

	USA	OOECD	USSR	CHINA	ROW
GDPO	5.6	10.2	2.68	1.1	3.34
KGDP	2.4	2.8	3.0	3.0	3.0
DEPR	5.00	5.00	5.00	5.00	5.00
PNREF	3.50	4.00	2.00	2.00	2.00
ESUB	0.40	0.40	0.30	0.30	0.30
KPVS	0.24	0.28	0.30	0.30	0.30
ELVS	0.33	0.33	0.33	0.33	0.33

* MAXIMUM DECLINE AND EXPANSION FACTORS, ANNUAL.

	USA	OOECD	USSR	CHINA	ROW
DECF	0.95	0.95	0.95	0.95	0.95
EXPF	1.1487	1.1487	1.1487	1.1487	1.1487
NXPF	1.1487	1.1487	1.1487	1.1487	1.1487

* MISCELLANEOUS FACTORS.

	USA	OOECD	USSR	CHINA	ROW
OGPD	1.25	1.25	1.25	1.25	1.25
CLGDP					

Terminology:

GDPO   INITIAL GDP ($ TRILLIONS):
KGDP   INITIAL CAPITAL-GDP RATIO
DEPR   ANNUAL PERCENT DEPRECIATION

PNREF REFERENCE PRICE OF NON-ELECTRIC ENERGY - DOLLARS PER GJ
ESUB   ELASTICITY BETWEEN K-L AND E-N
KPVS   CAPITAL VALUE SHARE PARAMETER
ELVS   ELECTRIC VALUE SHARE PARAMETER
AEEI   AUTONOMOUS ENERGY EFFICIENCY IMPROVEMENT - PERCENT PER YEAR

DECF   MAX DECLINE FACTOR (ANNUAL) FOR DL TECHNOLOGIES
EXPF   MAX EXPANSION FACTOR (ANNUAL) FOR EX (ELECTRIC) TECHNOLOGIES
NXPF   MAX EXPANSION FACTOR (ANNUAL) FOR NX (NONELECTRIC) TECHNOLOGIES

OGPD   OIL-GAS PRICE DIFFERENTIAL - DOLLARS PER GJ
CLGDP COAL-GDP GROWTH ELASTICITY

TABLE GROW(*, RG) POTENTIAL GDP GROWTH RATES - ANNUAL PERCENT

	USA	OOECD	USSR	CHINA	ROW
1990	2.50	2.70	2.50	4.50	3.75
2000	2.00	2.00	2.00	4.00	3.30
2010	2.00	2.00	2.00	4.00	3.30
2020	1.75	1.75	1.75	3.75	3.05
2030	1.50	1.50	1.50	3.50	2.80
2040	1.50	1.50	1.50	3.50	2.80
2050	1.25	1.25	1.25	3.25	2.55
2060	1.25	1.25	1.25	3.25	2.55
2070	1.125	1.125	1.125	3.125	2.425
2080	1.00	1.00	1.00	3.00	2.30
2090	1.00	1.00	1.00	3.00	2.30
2100	1.00	1.00	1.00	3.00	2.30

TABLE AEEI(*,RG) AUTONOMOUS ENERGY EFFICIENCY IMPROVEMENT - PERCENT PER YEAR

* SHOULD NOT EXCEED MACROECONOMIC GROWTH RATE.

	USA	OOECD	USSR	CHINA	ROW
1990	.50	.50	.25	1.00	.00
2000	.50	.50	.25	1.00	.00
2010	.50	.50	.30	.90	.10
2020	.50	.50	.35	.80	.20
2030	.50	.50	.40	.70	.30
2040	.50	.50	.45	.60	.40
2050	.50	.50	.50	.50	.50
2060	.50	.50	.50	.50	.50
2070	.50	.50	.50	.50	.50
2080	.50	.50	.50	.50	.50
2090	.50	.50	.50	.50	.50
2100	.50	.50	.50	.50	.50

Appendix B

Electric Power Technologies (ELEC.TAB)

TABLE   ECAP(*, RG, ET) ELECTRICITY PRODUCTION CAPACITIES - TKWH

	HYDRO	GAS-R	GAS-N	OIL-R
1990.USA	.190	.270		.160
2000.USA	.190	.135	.2	.080
2010.USA	.190		.8	
2020.USA	.190		3.2	
2030.USA	.190		6.4	
2040.USA	.190		12.8	
2050.USA	.190		25.6	
2060.USA	.190		51.2	
2070.USA	.190		100.0	
2080.USA	.190		100.0	
2090.USA	.190		100.0	
2100.USA	.190		100.0	

*	HYDRO	GAS-R	GAS-N	OIL-R
1990.OOECD	.835	.290		.400
2000.OOECD	.835	.145	.2	.200
2010.OOECD	.835		.8	
2020.OOECD	.835		3.2	
2030.OOECD	.835		6.4	
2040.OOECD	.835		12.8	
2050.OOECD	.835		25.6	
2060.OOECD	.835		51.2	
2070.OOECD	.835		100.0	
2080.OOECD	.835		100.0	
2090.OOECD	.835		100.0	
2100.OOECD	.835		100.0	

*	HYDRO	GAS-R	GAS-N	OIL-R
1990.USSR	.213	.566		.227
2000.USSR	.213	.283	.2	.113
2010.USSR	.213		.8	
2020.USSR	.213		3.2	
2030.USSR	.213		6.4	
2040.USSR	.213		12.8	
2050.USSR	.213		25.6	
2060.USSR	.213		51.2	
2070.USSR	.213		100.0	
2080.USSR	.213		100.0	
2090.USSR	.213		100.0	
2100.USSR	.213		100.0	

*	HYDRO	GAS-R	GAS-N	OIL-R
1990.CHINA	.099	.001		.059
2000.CHINA	.099	.0005	.2	.030
2010.CHINA	.099		.8	
2020.CHINA	.099		3.2	
2030.CHINA	.099		6.4	
2040.CHINA	.099		12.8	
2050.CHINA	.099		25.6	
2060.CHINA	.099		51.2	
2070.CHINA	.099		100.0	
2080.CHINA	.099		100.0	
2090.CHINA	.099		100.0	
2100.CHINA	.099		100.0	

*	HYDRO	GAS-R	GAS-N	OIL-R
1990.ROW	.452	.400		.400
2000.ROW	.452	.200	.2	.200
2010.ROW	.452		.8	
2020.ROW	.452		3.2	
2030.ROW	.452		6.4	
2040.ROW	.452		12.8	
2050.ROW	.452		25.6	
2060.ROW	.452		51.2	
2070.ROW	.452		100.0	
2080.ROW	.452		100.0	
2090.ROW	.452		100.0	
2100.ROW	.452		100.0	

+	COAL-R	COAL-N	NUC-R	ADV-HC	ADV-LC
1990.USA	1.479		.562		
2000.USA	1.479	100.0	.562		
2010.USA	1.109	100.0	.562	.1	
2020.USA	0.740	100.0	.280	100.0	.1
2030.USA	0.370	100.0		100.0	100.0
2040.USA		100.0		100.0	100.0
2050.USA		100.0		100.0	100.0
2060.USA		100.0		100.0	100.0
2070.USA		100.0		100.0	100.0
2080.USA		100.0		100.0	100.0
2090.USA		100.0		100.0	100.0
2100.USA		100.0		100.0	100.0

*	COAL-R	COAL-N	NUC-R	ADV-HC	ADV-LC
1990.OOECD	.916		.847		
2000.OOECD	.916	100.0	.847		
2010.OOECD	.687	100.0	.847	.2	
2020.OOECD	.458	100.0	.424	100.0	.2
2030.OOECD	.229	100.0		100.0	100.0
2040.OOECD		100.0		100.0	100.0
2050.OOECD		100.0		100.0	100.0
2060.OOECD		100.0		100.0	100.0
2070.OOECD		100.0		100.0	100.0
2080.OOECD		100.0		100.0	100.0
2090.OOECD		100.0		100.0	100.0
2100.OOECD		100.0		100.0	100.0

*	COAL-R	COAL-N	NUC-R	ADV-HC	ADV-LC
1990.USSR	.191		.209		
2000.USSR	.191	100.0	.209		
2010.USSR	.143	100.0	.157	.1	
2020.USSR	.096	100.0	.105	100.0	.1
2030.USSR	.048	100.0	.052	100.0	100.0
2040.USSR		100.0		100.0	100.0
2050.USSR		100.0		100.0	100.0
2060.USSR		100.0		100.0	100.0
2070.USSR		100.0		100.0	100.0
2080.USSR		100.0		100.0	100.0
2090.USSR		100.0		100.0	100.0
2100.USSR		100.0		100.0	100.0

*	COAL-R	COAL-N	NUC-R	ADV-HC	ADV-LC
1990.CHINA	.382				
2000.CHINA	.382	100.0			
2010.CHINA	.287	100.0		.1	
2020.CHINA	.191	100.0		100.0	.1
2030.CHINA	.096	100.0		100.0	100.0
2040.CHINA		100.0		100.0	100.0
2050.CHINA		100.0		100.0	100.0
2060.CHINA		100.0		100.0	100.0
2070.CHINA		100.0		100.0	100.0
2080.CHINA		100.0		100.0	100.0
2090.CHINA		100.0		100.0	100.0
2100.CHINA		100.0		100.0	100.0

*	COAL-R	COAL-N	NUC-R	ADV-HC	ADV-LC
1990.ROW	.859				
2000.ROW	.859	100.0			
2010.ROW	.644	100.0		.1	
2020.ROW	.430	100.0		100.0	.1
2030.ROW	.215	100.0		100.0	100.0
2040.ROW		100.0		100.0	100.0
2050.ROW		100.0		100.0	100.0
2060.ROW		100.0		100.0	100.0
2070.ROW		100.0		100.0	100.0
2080.ROW		100.0		100.0	100.0
2090.ROW		100.0		100.0	100.0
2100.ROW		100.0		100.0	100.0

TABLE   ECST(RG,ET) ELECTRICITY COST COEFFICIENTS - MILLS PER KWH

	HYDRO	GAS-R	GAS-N	OIL-R	NUC-R
USA	2.6	3.2	13.7	4.3	20.6
OOECD	2.6	3.2	13.7	4.3	20.6
USSR	2.6	3.2	13.7	4.3	20.6
CHINA	2.1	2.6	13.1	3.4	18.2
ROW	3.6	4.5	15.0	6.0	25.4

+	COAL-R	COAL-N	ADV-HC	ADV-LC
USA	20.1	51.0	75.0	50.0
OOECD	23.2	51.0	75.0	50.0
USSR	20.1	51.0	75.0	50.0
CHINA	26.1	51.0	75.0	50.0
ROW	27.9	51.0	75.0	50.0

HYDRO: hydroelectric, geothermal and other existing low-cost renewables. No significant expansion after 1990.

GAS-R and OIL-R refer, respectively, to existing oil and gas-fired plants. Capital charges are sunk costs and therefore excluded from this analysis. To avoid double counting of fuel inputs, their cost coefficients include only O&M. Other activities include fuel, O&M and levelized capital costs.

GAS-N: advanced combined cycle.

COAL-R: coal - remaining
COAL-N: coal - new

NUC-R: nuclear - remaining

ADV-LC: advanced - low cost
ADV-HC: advanced - high cost

TABLE NCAP(*, RG, NT) NONELECTRIC PRODUCTION CAPACITIES - EXAJ

	GAS-LC	GAS-HC	OIL-LC	OIL-HC	OIL-MX	CLDU	SYNF	RNEW	NE-BAK
1990.USA	18.44		17.39		17.71	2.74			
2000.USA	1000	1000	1000	1000				5	
2010.USA	1000	1000	1000	1000			1	10	1
2020.USA	1000	1000	1000	1000			1000	10	1000
2030.USA	1000	1000	1000	1000			1000	10	1000
2040.USA	1000	1000	1000	1000			1000	10	1000
2050.USA	1000	1000	1000	1000			1000	10	1000
2060.USA	1000	1000	1000	1000			1000	10	1000
2070.USA	1000	1000	1000	1000			1000	10	1000
2080.USA	1000	1000	1000	1000			1000	10	1000
2090.USA	1000	1000	1000	1000			1000	10	1000
2100.USA	1000	1000	1000	1000			1000	10	1000

*	GAS-LC	GAS-HC	OIL-LC	OIL-HC	OIL-MX	CLDU	SYNF	RNEW	NE-BAK
1990.OOECD	13.59		13.59		26.57	5.25			
2000.OOECD	1000	1000	1000	1000				5	
2010.OOECD	1000	1000	1000	1000			1	10	1
2020.OOECD	1000	1000	1000	1000			1000	10	1000
2030.OOECD	1000	1000	1000	1000			1000	10	1000
2040.OOECD	1000	1000	1000	1000			1000	10	1000
2050.OOECD	1000	1000	1000	1000			1000	10	1000
2060.OOECD	1000	1000	1000	1000			1000	10	1000
2070.OOECD	1000	1000	1000	1000			1000	10	1000
2080.OOECD	1000	1000	1000	1000			1000	10	1000
2090.OOECD	1000	1000	1000	1000			1000	10	1000
2100.OOECD	1000	1000	1000	1000			1000	10	1000

*	GAS-LC	GAS-HC	OIL-LC	OIL-HC	OIL-MX	CLDU	SYNF	RNEW	NE-BAK
1990.USSR	26.29		24.87		-6.50	11.44			
2000.USSR	1000	1000	1000	1000				5	
2010.USSR	1000	1000	1000	1000			1	10	1
2020.USSR	1000	1000	1000	1000			1000	10	1000
2030.USSR	1000	1000	1000	1000			1000	10	1000
2040.USSR	1000	1000	1000	1000			1000	10	1000
2050.USSR	1000	1000	1000	1000			1000	10	1000
2060.USSR	1000	1000	1000	1000			1000	10	1000
2070.USSR	1000	1000	1000	1000			1000	10	1000
2080.USSR	1000	1000	1000	1000			1000	10	1000
2090.USSR	1000	1000	1000	1000			1000	10	1000
2100.USSR	1000	1000	1000	1000			1000	10	1000

*	GAS-LC	GAS-HC	OIL-LC	OIL-HC	OIL-MX	CLDU	SYNF	RNEW	NE-BAK
1990.CHINA	.53		5.74		-0.56	17.59			
2000.CHINA	1000	1000	1000	1000				5	
2010.CHINA	1000	1000	1000	1000			1	10	1
2020.CHINA	1000	1000	1000	1000			1000	10	1000
2030.CHINA	1000	1000	1000	1000			1000	10	1000
2040.CHINA	1000	1000	1000	1000			1000	10	1000
2050.CHINA	1000	1000	1000	1000			1000	10	1000
2060.CHINA	1000	1000	1000	1000			1000	10	1000
2070.CHINA	1000	1000	1000	1000			1000	10	1000
2080.CHINA	1000	1000	1000	1000			1000	10	1000
2090.CHINA	1000	1000	1000	1000			1000	10	1000
2100.CHINA	1000	1000	1000	1000			1000	10	1000

*	GAS-LC	GAS-HC	OIL-LC	OIL-HC	OIL-MX	CLDU	SYNF	RNEW	NE-BAK
1990.ROW	14.04		72.59		-37.22	12.23			
2000.ROW	1000	1000	1000	1000				5	
2010.ROW	1000	1000	1000	1000			1	10	1
2020.ROW	1000	1000	1000	1000			1000	10	1000
2030.ROW	1000	1000	1000	1000			1000	10	1000
2040.ROW	1000	1000	1000	1000			1000	10	1000
2050.ROW	1000	1000	1000	1000			1000	10	1000
2060.ROW	1000	1000	1000	1000			1000	10	1000
2070.ROW	1000	1000	1000	1000			1000	10	1000
2080.ROW	1000	1000	1000	1000			1000	10	1000
2090.ROW	1000	1000	1000	1000			1000	10	1000
2100.ROW	1000	1000	1000	1000			1000	10	1000

To allow for process heat produced within the electric power sector, have transferred all heat demands to CLDU within USSR and ROW.

The gas resource base may include pipeline and LNG imports as well as domestic production.

In the case of low-cost exhaustible resources, the production-reserve ratio is determined by the actual values for 1990; for high-cost exhaustible resources, by the PRV factor indicated in SDAT.

Costs for the CLDU activity (direct uses of coal) includes not only fuel, but also the differential in utilization costs between coal and oil/gas-fired units.

Within ROW, the SYNF activity includes unconventional oils such as tar sands and heavy oils.

TABLE NCST(RG, NT) NONELECTRIC COST COEFFICIENTS - $ PER GJ

	CLDU	SYNF	RNEW	NE-BAK	GAS-LC	GAS-HC	OIL-LC	OIL-HC
USA	2.0	8.333	6.0	16.667	1.5	5.0	2.5	6.0
OOECD	3.0	8.333	6.0	16.667	1.5	5.0	2.5	6.0
USSR	2.0	8.333	6.0	16.667	1.5	5.0	2.5	6.0
CHINA	2.0	8.333	6.0	16.667	1.5	5.0	2.5	6.0
ROW	2.0	8.333	6.0	16.667	0.5	5.0	1.0	6.0

TABLE SDAT(EI,X,*)   SUPPLY DATA - EXHAUSTIBLE HYDROCARBON RESOURCES

	USA	OOECD	USSR	CHINA	ROW	WORLD
RDF.OIL-LC	.05	.03	.03	.03	.10	
RSV.OIL-LC	361	294	612	147	3812	5226
RSC.OIL-LC	248	364	627	294	1496	3029
RDF.OIL-HC	.03	.03	.03	.03	.10	
PRV.OIL-HC	.05	.05	.05	.05	.05	
RSC.OIL-HC	248	364	627	294	1496	3029
RDF.GAS-LC	.05	.03	.03	.03	.10	
RSV.GAS-LC	352	476	1372	27	1979	4206
RSC.GAS-LC	255	628	1540	252	1950	4625
RDF.GAS-HC	.03	.03	.03	.03	.10	
PRV.GAS-HC	.05	.03	.03	.03	.03	
RSC.GAS-HC	255	628	1540	252	1950	4625

   RSV and RSC are base year levels of low-cost reserves and resources (exaj).
   PRV is the production-reserve ratio for high-cost resources. That for
low-cost resources is computed from base-year data.
   RDF measures fraction of remaining resources depleted and converted annually
into proven reserves.
   Undiscovered resources are split equally between low- and high-cost
categories.

SET

SO   INTERNATIONAL OIL PRICE SCENARIOS
     /S1/

TABLE OILP(*,*) INTERNATIONAL OIL PRICES (DOLLARS PER GJ)

* Caveat: Base year (1990) price must be identical in all scenarios.

	S1
1990	4.00
2000	5.10
2010	6.20
2020	7.30
2030	9.00
2040	8.40
2050	8.40
2060	8.40
2070	8.40
2080	8.40
2090	8.40
2100	8.40

TABLE OILX(*,*) OIL EXPORT LIMITS (EXAJ)

	USA	OOECD	USSR	CHINA	ROW
1990			6.6	.6	1000
2000			6.6	.6	1000
2010			5.0	.6	1000
2020			3.0	.6	1000
2030			1.0	.6	1000
2040					1000
2050					1000
2060					1000
2070					1000
2080					1000
2090					1000
2100					1000

The variable PN("OIL-MX") represents imports - exports of oil. If
the value of this variable is positive, then oil is being imported. Negative
values imply oil exports. We use the OIL* tables to set bounds on
PN("OIL-MX"). If there are limits on imports, then PN("OIL-MX") has an upper
bound equal to the value in the table OILM. If there are limits on exports,
then PN("OIL-MX") is given a lower bound equal to the negative of the value in
the OILX table.

The user must check to ensure that the export limits for USSR and CHINA
are logically consistent with the import limits of the OECD region. E.g., if
the OECD'S import limits are zero, the export limits for USSR and CHINA should
also be zero.

TABLE OILM(*,RG) OIL IMPORT LIMITS (EXAJ)

	USA	OOECD	USSR	CHINA	ROW
1990	1000	1000			
2000	1000	1000			
2010	1000	1000			
2020	1000	1000			
2030	1000	1000			
2040					
2050					
2060					
2070					
2080					
2090					
2100					

Export limits for ROW are calculated as a residual: imports by
USA + OOECD minus exports by USSR + CHINA. For this reason, the ROW model
must be calculated after the other four regions have performed their
individual optimizations. If ROW export limits are non-binding, there is an
unexplained global gap - an excess of demands over supplies.

Appendix E

Carbon Emissions

(CARBON.TAB)

```
SCALAR
         MXDIF    CARBON TAX - IMPORT - EXPORT VALUE DIFFERENTIAL        /20/
SET

         SC   INTERNATIONAL CARBON EMISSION PRICE SCENARIOS
              / S1/

TABLE CARP(*,*) INTERNATIONAL CARBON PRICES (DOLLARS PER TON)

*  Caveat: Base year (1990) price and quantity limits must be zero in all
*  scenarios.

              S1

*      Values from 2000-2050 transferred from MINDIST, 4/5/91.

2000        106.9745
2010        117.7829
2020        180.8409
2030        130.3588
2040        188.9276
2050        211.3755

2060        208.35
2070        208.35
2080        208.35
2090        208.35
2100        208.35

TABLE CARLIM(*, *) ANNUAL CARBON LIMITS  - BILLION TONS

* 20% REDUCTION BY 2010 IN USA, OOECD, USSR;
* 50% INCREASE IN CHINA,ROW.
```

	USA	OOECD	USSR	CHINA	ROW	WORLD
1990	1.430	1.375	1.055	.641	1.502	6.003
2000	1.430	1.375	1.055	.801	1.878	6.539
2010	1.144	1.100	.844	.961	2.254	6.303
2020	1.144	1.100	.844	.961	2.254	6.303
2030	1.144	1.100	.844	.961	2.254	6.303
2040	1.144	1.100	.844	.961	2.254	6.303
2050	1.144	1.100	.844	.961	2.254	6.303
2060	1.144	1.100	.844	.961	2.254	6.303
2070	1.144	1.100	.844	.961	2.254	6.303
2080	1.144	1.100	.844	.961	2.254	6.303
2090	1.144	1.100	.844	.961	2.254	6.303
2100	1.144	1.100	.844	.961	2.254	6.303

```
$ONTEXT
```

The values of the following carbon emission coefficients are based upon
EMF-12 ground rules:

```
    oil:  .01994 billion metric tons/exaj
    gas:  .01374 "                      "
    coal: .02412 "                      "
    synf: .04000 "                      "
```

Electric coefficients are expressed per TWH. Heat rates for coal-fired
plants are multiplied by carbon coefficient shown above. Avoid double counting
by setting carbon emission coefficients equal to zero for oil and gas-fired
plants.

```
$ OFFTEXT
```

TABLE CH(AT, CD, RG) CARBON EMISSION COEFFICIENTS AND HEAT RATES

	USA	OOECD	USSR	CHINA	ROW
COAL-R.CEC	.2779	.2844	.2680	.2776	.3600
COAL-N.CEC	.2533	.2533	.2533	.2533	.2533
GAS-LC.CEC	.01374	.01374	.01374	.01374	.01374
GAS-HC.CEC	.01374	.01374	.01374	.01374	.01374
OIL-LC.CEC	.01994	.01994	.01994	.01994	.01994
OIL-HC.CEC	.01994	.01994	.01994	.01994	.01994
OIL-MX.CEC	.01994	.01994	.01994	.01994	.01994
CLDU.CEC	.02412	.02412	.02412	.02412	.02412
SYNF.CEC	.04000	.04000	.04000	.04000	.04000

	USA	OOECD	USSR	CHINA	ROW
HYDRO.HTRT	11.5	11.8	11.5	11.5	14.5
OIL-R.HTRT	11.5	11.8	11.5	11.5	14.5
GAS-R.HTRT	11.5	11.8	11.1	11.5	14.925
COAL-R.HTRT	11.52	11.79	11.5	11.51	14.5
NUC-R.HTRT	11.5	11.8	11.5	11.5	14.5
GAS-N.HTRT	8.0	8.0	8.0	8.0	8.0
COAL-N.HTRT	10.5	10.5	10.5	10.5	10.5
ADV-LC.HTRT	10.5	10.5	10.5	10.5	10.5
ADV-HC.HTRT	10.5	10.5	10.5	10.5	10.5

```
SCALAR   SYNTPE   ADJUSTMENT FOR SYNFUELS IN TPE REPORT    /.66/;

*  SYNTPE = (SYNF.CEC / CLDU.CEC) - 1.
```

TABLE CARXL(*,*) CARBON EXPORT LIMITS (BILLION TONS)

	USA	OOECD	USSR	CHINA	ROW
1990					
2000	1000	1000	1000	1000	1000
2010	1000	1000	1000	1000	1000
2020	1000	1000	1000	1000	1000
2030	1000	1000	1000	1000	1000
2040	1000	1000	1000	1000	1000
2050	1000	1000	1000	1000	1000
2060	1000	1000	1000	1000	1000
2070	1000	1000	1000	1000	1000
2080	1000	1000	1000	1000	1000
2090	1000	1000	1000	1000	1000
2100	1000	1000	1000	1000	1000

TABLE CARML(*,*) CARBON IMPORT LIMITS (BILLION TONS)

	USA	OOECD	USSR	CHINA	ROW
1990					
2000	1000	1000	1000	1000	1000
2010	1000	1000	1000	1000	1000
2020	1000	1000	1000	1000	1000
2030	1000	1000	1000	1000	1000
2040	1000	1000	1000	1000	1000
2050	1000	1000	1000	1000	1000
2060	1000	1000	1000	1000	1000
2070	1000	1000	1000	1000	1000
2080	1000	1000	1000	1000	1000
2090	1000	1000	1000	1000	1000
2100	1000	1000	1000	1000	1000

Index